The Naked Emperor

The Naked Emperor:
Darwinism Exposed

Antony Latham

JANUS PUBLISHING COMPANY
London, England

First Published in Great Britain 2005
by Janus Publishing Company Ltd,
105-107 Gloucester Place,
London W1U 6BY

www.januspublishing.co.uk

British Library Cataloguing-in-Publication Data
A catalogue record for this book
is available from the British Library

ISBN 1 85756 635 1

Cover Design Simon Hughes and Antony Latham

Front cover image of Cyanobacterium (the first type of life known)
© Dennis Kunkel Microscopy, Inc.

Printed and bound in Great Britain

This book is lovingly dedicated to my wife, Christian,
who has put up with a distracted husband for far too long.

The Windhover[1]

To Christ our Lord

I caught this morning morning's minion, king-
  dom of daylight's dauphin, dapple-dawn-drawn Falcon, in
     his riding
  Of the rolling level underneath him steady air, and striding
High there, how he rung upon the rein of a wimpling wing
In his ecstacy! then off, off forth on swing,
  As a skate's heel sweeps smooth on a bow-bend: the hurl
     and gliding
  Rebuffed the big wind. My heart in hiding
Stirred for a bird, - the achieve of, the mastery of the thing!

Brute beauty and valour and act, oh, air, pride, plume, here
  Buckle! AND the fire that breaks from thee then, a billion
Times told lovelier, more dangerous, O my chevalier!

  No wonder of it: sheer plod makes plough down sillion
Shine, and blue-bleak embers, ah my dear,
  Fall, gall themselves, and gash gold-vermillion.

Gerard Manley Hopkins

1 Windhover is old English for Kestrel.

Contents

Acknowledgements

I am immensely grateful to all those who have read the early manuscript and given advice and encouragement. I particularly want to thank Catherine Davies, James Finlayson, James Hardy and John Latham.

Many thanks to Denis Alexander who, though not sharing my views about evolution, gave me help with the early chapters. I am indebted to Simon Conway Morris, who also comes to different conclusions from me but nevertheless spent much time going through the entire manuscript with his geological hammer, giving invaluable comments and correction.

Thanks also to Sandy Leung, Managing Director of Janus Publishing who had faith in this book from the beginning and I am particularly grateful to Simon Hughes, Creative Director at Janus, who has been constantly helpful in the production and design.

Any errors or faults in the book are of course entirely my own.

Preface

I am writing this book as a very personal journey, one however, which is relevant to any sceptical reader seeking the truth about our origins. It is a journey concerned with a huge question: how did we, as human beings, come to be?

I am convinced that people are still fascinated by this matter. This is because we all base our lives, even if unconsciously, on how we see our place in the cosmos.

Since the general acceptance of Darwin's theory of evolution, mankind has had to deal with the appalling possibility that we are here by accident. If I fully accept evolutionary theory then I cannot believe that there is any ultimate purpose to my life – nor can I believe in a creator God.

Many believers in God will dispute this and assert that they can cope quite well with a God-driven evolution. I will attempt to show that they are mistaken if they believe this alongside classical Darwinism.

My journey starts in biology class in secondary school, where I learnt the rudiments of evolutionary theory. I was, I think, just 13 years old and I loved the biology lessons. They were given by a dedicated and gentle teacher who loved his subject. We were surrounded by the wonderful paraphernalia of a good biology classroom; fish tanks, skeletons, specimens in formalin jars and breeding locusts. To me the whole atmosphere was charged with the glory of nature.

We followed the standard teaching about evolution. I learnt about how natural selection 'chooses' the fittest amongst the variants thrown up, by chance, amongst organisms. The fittest survived and so were selected to carry on and perpetuate their genes. I soon understood that the process of producing variation was ultimately that of mutations. Mutations were random 'mistakes' in the genetic code and sometimes the mistakes produced new, fitter forms.

It did not hit me immediately but I soon realised that the rug of religion had been well and truly pulled from under my feet. It did not worry me too much then; I just thought I had grown up and need not see any requirement for God. If he did not create us (and random mutations seemed to prove he didn't) then he may as well not exist. Until then, in my childish way, I had assumed that the wonders of nature were created and beautifully designed. My biology teacher had told me otherwise.

The fact that he was also a lay-preacher in the church just seemed to confirm to me how out of touch with reality Christianity was. Christianity, as I knew it, consisted of dry and boring sermons and rather meaningless ceremonies.

The exception to this jaundiced view of Christianity was in my attitude to the person of Jesus. Jesus, continued, as taught to me by my mother, to be a truly wonderful person. I never let go of this concept of Jesus but certainly doubted if he was the Son of God.

So, as I progressed through my teens, university and life as a doctor, I was a supreme sceptic as far as religion was concerned. I enjoyed disputing with Christians about evolution. I admired their faith but felt it was groundless. Invariably I would bring up the 'fact' of evolution. To me it was simply a knock-down argument against belief.

I well remember a popular series on television which I watched as a teenager. It was a dramatised account of the life of Darwin. I was fascinated – particularly during the part about his voyage around the world on HMS *Beagle*. I loved the quiet and methodical way in which he put all the facts together. I was very struck by the evidence he found on the Galapagos Islands. He discovered there the various species of finches which had so clearly evolved on the different islands – all from a common ancestor. I admired the way in which he struggled with his theory for many years before being forced to publish it. I liked his overall humble disposition, the epitome of the best kind of scientist.

Rarely did any news of creationists filter into my world-view. Whenever I did encounter creationist beliefs I would inwardly scoff. Such ideas, it seemed to me, belonged to the Middle Ages and were akin to belief in a flat earth. Nevertheless, some part of me admired anyone who clung to their faith despite (what seemed to me) incontrovertible evidence to the contrary.

In 1982 I was in my third year working as a doctor in a remote hospital in north-west Kenya. I became a Christian there. Given my previous scepticism this was an astounding conversion. For days afterwards I marvelled that it was not an illusion. It wasn't. God had invaded my life despite my views on nature.

I do not want here to go into how or why I was converted. What I can say is that the experience had profound implications for how I viewed nature. Nature was no longer a random accident but an expression of God, who created everything.

I knew that I would have to return sometime to the whole creation question and see whether I had got it wrong beforehand. In my heart I knew that we were created – it was just that my head, and all its knowledge about biology, had not caught up.

And so as a Christian I had a tension between my faith and my understanding of biology; this is why many years later, I was pleased to have, at last, the time to really look hard at the evidence.

My family and I had returned from working in Tanzania at the end of 1995 and I began working as a general practitioner in Scotland. When I arrived back in this culture I determined to do plenty of reading around the subject of evolution. I started, somewhat fearfully, by reading a book by the well-known biologist Richard Dawkins. I had read reviews of his books and knew that he was a fervent atheist and a professor in Oxford. I had read that his books were seen as attacks upon religious belief, based on the evidence for evolution. A sizeable section of the population and the scientific establishment revered him.

My aim, as always, was to look as objectively as possible at the known facts. I have the belief that God's world is what science investigates and so, as a Christian, I had nothing to fear in looking into that world with open scientific eyes.

I was expecting Dawkins' book to shake me to the core. His reputation had gone before him. I braced myself to be confronted with devastating, razor-sharp intelligence, upholding his atheist stance.

Well (and I'm sorry about the anti-climax) – he disappointed me. The book which I read, called *River Out of Eden*,[1] contained nothing whatever to prove his point. Readers may feel that I am arrogant for writing this. However, I read his book (and others since) very methodically, more than once. Nothing in it seemed particularly scientific to me because all his evolutionary scenarios were quite clearly based on his prior beliefs. His method of describing the processes of evolution was to read into the wonders of nature what he already believed. There seemed nothing very logical about this. Nowhere could I pin down a single argument of his and describe it as one would, say, Newton's laws of gravity.

It reminded me of the story of the emperor's clothes. Everyone was crowing on about how wonderful they were but careful review showed them to be entirely absent. So, it seemed to me, were the arguments of Richard Dawkins. I will amplify what I mean about those emperor's clothes later (the last section of this book is a blow-by-blow critique of his book *The Blind Watchmaker*)', but having managed to read Dawkins and kept my faith intact, I was ready for some serious investigation. Some might feel that I have some sort of vendetta against Richard Dawkins; nothing is further from the truth. I have nothing personal against him; it is simply that he has made himself the principal apologist for Darwinism for our generation and must therefore be taken very seriously.

Thankfully I have been helped along the way by many good books which support evidence for an Intelligent Designer of life – not least the classic *'Darwin on Trial'* by Phillip Johnson.[2]

Inevitably I have had to delve deep into matters of evolutionary theory. A few sections of this book therefore are fairly technical and I have mentioned these and sectioned them off in italics for those who would prefer to skip them. I would encourage you to at least skim read these bits so that you can get a flavour of the details.

This is a fascinating journey and I have not finished it. Do not expect cut and dried sewn-up arguments. Expect instead a serious (and hopefully enjoyable) critique of Darwinism with some unanswered questions. I have looked long and hard at the empirical scientific evidence for evolution in the areas of cosmology, palaeontology, genetics, biochemistry, anthropology and psychology. There is very little here that is original; I have gleaned from others and have acknowledged this in the footnotes and bibliography. You will find that I come to a clear conclusion that the appearance of life in all its forms is no accident.

A crucial point, worth emphasising now, is that there is a world of difference between micro-evolution and macro-evolution. This will be explained in the book – please be patient as you read.

Micro-evolution is a fact and is explicable on the basis of changing proportions of genes in different populations. Darwin based much of his theory on micro-evolution. As I shall explain later, there is in fact no new genetic information in micro-evolution and it can only produce small changes in organisms.

Macro-evolution, on the other hand, refers to the large-scale new structures and body plans that we see in nature. I will show that there is no explanation in Darwinism for these leaps of form.

You will quickly discern that I am not a 'young earth' creationist. If that upsets you then you may wish to stop reading now. However, I would strongly encourage you to look at the reasoning for this and, if you are a Bible believer, come to the conclusion with me that an ancient earth is compatible with the Genesis story. I am not going to discuss the theology behind this stance here – there are other books which do this.

This book is not primarily for Christians. It is for anyone who genuinely wants another scientific view on nature.

I am rather tired with the bland and patronising statements about evolution which we receive regularly from various experts in the media. I believe strongly that the story of how we got here is the property of the human race – not of a few scientists. I want here to de-mystify some of the facts, take them out of the ivory towers, and allow you, the reader, to make your own mind up.

1. Phillip Johnson (1991). Darwin on Trial. Washington DC: Regnery Gateway.
2. Richard Dawkins (1995). River Out of Eden London: Weidenfeld & Nicolson.

Chapter One

A Fine-tuned Universe

Any study of the processes that have resulted in life on earth cannot avoid a serious look at what we understand about the universe. There must be some continuity between the universe in which our fragile planet resides and the astonishing profusion of life with which we are blessed.

The universe is very finely tuned. This is now generally accepted amongst cosmologists. The particular conditions that existed at the big bang 13 billion years ago had to be very precise in order to end up with the sort of 'biophilic' cosmos that can allow for life. Another word to describe the universe is 'anthropic', meaning that the universe seems perfectly ordered to produce not just life but man. The conditions and laws embedded in those first few moments were critical. Tiny variations in any of the various conditions would have spelt the death of any such biophilic universe. To achieve the formation of galaxies, stars, our solar system and earth required such fine-tuning that one inevitably looks for the tuner.

It could have been otherwise. The various factors which were so delicately balanced at that primordial moment could theoretically have been quite different in endless ways. Why is the universe so well adjusted? Could it be that it is just one of many universes and we simply struck lucky? Or is it more likely to be a unique and designed system which is tailor-made for life? I will outline some of the factors that are so perfectly adjusted and then discuss the philosophical and theological issues surrounding them. I am thankful to Martin Rees, Astronomer Royal, for his book *Our Cosmic Habitat*, which has informed much of this chapter.[1]

When the big bang occurred there had to be a very slight unevenness in the expanding energy. If it had been entirely smooth or uniform then the universe would have no 'clumpiness' and there would never have been galaxies or stars. The amplitude of these initial non-uniformities can be described by a simple number, Q, which is the energy difference between the peaks and troughs in the density, expressed as a fraction of the total energy of the initial universe. Computer models show that Q had to be very close to 0.00001 in order for any galaxies to form. A slightly higher or lower value would mean no galaxies and no possibility of life. If it was very slightly smaller then no structures would have formed. If it was fractionally higher then all matter would have collapsed into huge black holes. The fact that Q needed to be so precise is just one of the evidences for design in our cosmos. Q could have had some other value. Why was it just right?

At the beginning of the universe there was matter and antimatter. If the amounts of each had been exactly the same then they theoretically should have cancelled each other out – leaving only energy. What is it that allowed some matter to survive and not be annihilated? The Russian physicist Andrei Sakharov showed that matter and antimatter are not precise mirror images of each other. There is a very slight asymmetry which favours matter over antimatter very early in the cosmic expansion. The difference is only about one part in a billion but this tiny difference is absolutely crucial. For every billion pairs of matter and antimatter that annihilate (yielding photons) just one unit of matter is left over. We and the rest of the material in the universe only exist because of this one in a billion difference. As Martin Rees writes: 'we owe our very existence to a difference in the ninth decimal place'.[2]

It is not just that there are atoms in the universe because of this slight asymmetry between matter and antimatter. The amount of atoms is also very important. (In fact atoms are not the dominant constituent of the universe. A mysterious substance called 'dark matter' outweighs all the atoms by a factor of five to ten.) But if there were, for example, ten times fewer atoms in the universe than there are, then they would remain in a diffuse gas that would never condense into galaxies and stars.

Compared to the forces within the atom, gravity is a very weak one. It is 10 to the power 36 times weaker than the electric force between two protons. Nevertheless it acts on large masses over great distances and the value of it is crucial to the survival of the universe as we know it. It is essential that the force of gravity be very weak. If it was stronger, then stars (including the sun) would be small and short lived, not allowing the time for life to flourish on the earth. Also, a much stronger gravity would crush any living thing that was larger than an insect. Because it is weak, gravity allows large objects in the universe to have very long lives and to be 'interesting' enough to produce life. As Martin Rees says: 'Although gravity is crucial in the cosmos, the weaker it is (provided it isn't zero) the grander and more complex can be its manifestations.'[3]

We now know that there is also a repulsive force in the universe, known as 'dark energy', which acts in the opposite way to gravity. Exactly what this is (or even what gravity is) is elusive, but it has been measured. Its force (known as lamda) is very weak also. If it had been large then this 'disruptive' force would have prevented gravitationally bound structures (such as the sun and planets) from forming.

That both gravity and dark energy are so weak is crucial. If they had other values then it is doubtful whether we would have the conditions for life that we need.

Staying with gravity; it was crucial for the expansion of the universe at the very first second of the big bang that the expansion energy (or impetus) was finely balanced with the gravitational force. If the expansion energy had been too great then galaxies and stars would never have been able to pull themselves together with gravity. If the expansion energy had been too small then there would have been a premature 'big crunch' as the universe imploded into itself. It has been mathematically calculated that, back at one second, the universe's kinetic (expansion) and gravitational energies must have differed by less that one part in a million billion (one in 10 to the power 15). Such an incredibly tiny difference between these two energies points to a designer who made it that way.

At the very micro level of atoms there is also extremely fine-tuning. There are two crucial forces within atoms: the electrical repulsion between protons, and the strong opposite nuclear force which attracts them to each other (and which also attracts the electrically neutral neutrons). If the nuclear forces were very slightly weaker then no chemical elements other than hydrogen would be stable. There would be no variety of elements and chemistry would be simply about hydrogen. There also would be no nuclear energy to power the stars. On the other hand, if the nuclear forces were fractionally stronger than they are relative to the electric forces, then two protons would stick together, ordinary hydrogen could not exist and stars such as our sun could not evolve.

We, and the rest of life, are made up of carbon-based chemistry. The carbon that is in you and me was manufactured in some star prior to the formation of our solar system. Without carbon there can be no life. Each carbon nucleus (six protons and six neutrons) is made from three nuclei of helium. Astrophysicists Hoyle and Salpeter worked out that this process of forming carbon works efficiently only if the carbon nucleus has a strange feature: a mode of vibration with a very specific energy. The vibration frequencies of atomic nuclei depend on the nuclear force (already described) that holds protons and neutrons together. If the force were changed by more than 1 or 2 per cent then carbon could not be formed. Once again we see very subtle tuning in the way things are.

A neutron is heavier than a proton by 0.14 per cent – a very small fraction. This difference, even though small, exceeds the total mass of an electron. This is crucial because if electrons were very minutely heavier then they would combine with protons to form neutrons – leaving no hydrogen and so no possibility of stars or chemistry as we know it. The masses of these subatomic particles give all the appearance of being designed to be just as they are.

It would be possible for me to catalogue other aspects of fine-tuning and to include the way in which our earth is not just any planet but seems to have been put in a position that seems very 'set up'. There is a very long list of pre-requisites for a biophilic life-sustaining earth that are almost certainly rare in the extreme: the orbit must not wander too close or too far away from the sun, its spin must be stable (something the moon ensures), there must not be excessive bombardment by asteroids – this is helped by the presence of Jupiter, which has a near-circular orbit that acts as a trap for stray asteroids. Its position in the Milky Way galaxy must be especially propitious to ensure that it is not irradiated heavily by cosmic rays or in danger of encounters with other stars.

We now have detected planets in orbit around distant stars and there must be billions of them in the universe. Perhaps there are other planets with the same profile as earth but they must be exceedingly rare.

Astronomer Hugh Ross has looked in detail at a long list of essential parameters that need to be right for life on planet earth.[4] These include all the fine-tuned factors I have mentioned as well as features such as the earth's axis tilt, rotation period, surface gravity, asteroidal collision rate and many others. Each parameter is given a probability and from this he calculates the probability of all 55 of those listed to be 10 to the power -69. It is estimated that the maximum number of planets in the universe is 10 to the power 22. This means that there is much less than one chance in one hundred billion trillion trillion trillion that another such planet could occur anywhere in the universe.

The fine-tuning of our universe and our planet earth is so perfect that it has been described as being like a pencil balancing on its point. It continues to balance, against all the odds, because the conditions are exactly set for this to happen.

Scientists have reacted in different ways to the evident fine-tuning that has been described. Some, such as John Polkinghorne, former Professor of Mathematical Physics at Cambridge University (and now an ordained theologian), see all of this as evidence of a God who has carefully and lovingly designed our universe.[5]

Others, such as Martin Rees, come to the interesting conclusion that there must be many universes. We, he says, are in just one of them which happens to have the appearance of fine-tuning, quite fortuitously. Many cosmologists favour this idea, even though there is absolutely no empirical data to support it. They favour it because it seems the only way out of ascribing the situation to God. If there are many, even an infinite number, of universes, then it is not surprising, they argue, if one of them, by chance, has the characteristics that ours has.

I find this argument for a 'multi-verse' to be very tenuous and forced. It goes against the grain on a philosophical level. There is a well-known principle in philosophy known as 'Occam's razor' or the 'principle of parsimony'. Essentially this principle states that when faced with theories about reality then the most simple or parsimonious is far more likely to be true. With the multi-verse theory, cosmologists are adding a complication for which there is no evidence and they are doing it for unscientific reasons. Paul Davies writes: 'Invoking an infinite number of other universes just to explain the apparent contrivances of the one we see is pretty drastic, and in stark conflict with Occam's razor.'[6] Physicist Edward Harrison writes: 'Take your choice: blind chance that requires multitudes of universes, or design that requires only one.'[7]

Pure science should not out rule something on the basis of whether it points to a creator or not. Pure and objective science will not twist and turn because of some inbuilt aversion to having God in the picture. The idea of multiple universes is the result of cosmologists unable to face a challenge to their own belief systems. It is interesting that Martin Rees, in his book *Our Cosmic Habitat* states that the multi-verse scenario is his preference without giving any explanation as to why this is so. Many other cosmologists however, have come to the conclusion that God is behind and in the details of our stunningly balanced and well-honed cosmos.

For me the fine-tuning described is the essential backdrop to the stage on which life came about. There is, however, a danger of deism here – a trap into which some theologians fall. A deist believes that God designed and set up the universe but that he has no active part in it now. Theism, in contrast, affirms the biblical stance that God is constantly active and involved in his creation and continues to uphold it. A theistic theology can then see no reason why God should not be active in the details of how animals and plants were formed on the earth. The deist will be happy to leave this to blind evolutionary chance. I hope that the rest of this book will give ample reason to believe in an active involvement by a designer in every stage of the process of life appearing.

1 Rees, Martin (2002). *Our Cosmic Habitat.* London: Weidenfeld & Nicolson.
2. Ibid.
3. Ibid.
4. See William A. Dembski (ed.) (1998). *Mere Creation. Science, Faith and Intelligent Design.* Inter Varsity Press.
5. Talk given by John Polkinghorne, BBC Radio 4, 2003.
6. Quote from: William A. Dembski 1998. *Mere Creation. Science, Faith & Intellegent design*
7. E. Harrison (1985). *Masks of the Universe.* New York: Collier Books, Macmillan.

Chapter Two

The First Life on earth

Our planet is 4.5 billion years old. This amount of time is virtually impossible for us to imagine but we must grapple with enormous periods of time as we pursue the trajectory of life. Even more difficult to conceive is the great length of time since the origin of the known universe – 13 billion years. And so, when the clouds of dust and gas which circled the sun coalesced to form the planets, the universe was already very old indeed.

The very early earth was a hot and hostile place. There was no solid crust on the surface. There was also continuous heavy bombardment from other stray 'bits' of the material circling the sun. It is thought that our moon may have formed from a large chunk of the earth, propelled into space by an enormous collision from a Mars-sized object.

For many millions of years, therefore, the earth was an unstable, molten inferno with little to commend it as a place for starting life. So when did life start? We now have a much clearer idea following the discovery of some extraordinary fossils in Australia in 1992.

One of the world's most renowned experts in micro-palaeontology (the study of microscopic fossils) is J. William Schopf of the University of California. He led a team to some rocks called the Apex Chert in a remote part of Western Australia in 1992. He found some fossil bacteria there.[1]

This in itself is nothing unusual; it was the age of these fossils which astonished every one. They were 3.46 billion years old.

The dating of these fossils is accomplished by measuring certain radioactive isotopes within the rocks. One of these, known as Uranium 238 decays, very slowly at a known rate, into Lead 205. By measuring the proportions of each, it is possible to give an accurate date for the formation of the rock concerned.

The fossils that he found are tiny single-celled bacteria. Not very exciting, you might be thinking. You would be wrong. These are tremendously exciting because amongst these bacteria can be identified some which are indistinguishable from a group of living bacteria called cyanobacteria. These are a form that use photosynthesis (as plants do) to harness the sun's energy to provide the power to live and grow. There is an electron micrograph of a Cyanobacterium cross-section on the front cover of this book.

These tiny fossil cells lie within strange formations known as cherts. These are associated with stromatolites. Stromatolites are mounds made by the bacteria, usually in shallow sea water in coastal areas. The bacteria make layer after layer of organic mats which, in association with sediment, eventually protrude out of the water. Today, living bacteria make stromatolites in a few places such as Western Australia and the Bahamas.

There is some evidence of even older bacteria in some ancient rocks in Isua, Greenland, aged 3.8 billion years. Geologists rarely find such old rocks because, before the time of their formation, the earth, as I have said, was in a molten flux with only a few crusts forming on the surface. They have found an unusual chemical signature within these rocks that points to life – life that was probably photosynthetic bacteria, like those found in the Apex Chert. It is some isotopes of the element carbon which give this picture. In these rocks there is a slight excess in the concentration of carbon 12 relative to carbon 13 and this is a characteristic signature of photosynthetic carbon-fixing.

Therefore, we have good evidence that complex photosynthetic bacteria were in existence 3.8 billion years ago. It is highly unlikely that any form of life would have been possible before 4.0

billion years ago and so we are looking here at complex life starting almost immediately, geologically speaking, after the conditions were right.

The samples of fossil bacteria found in the Apex Chert by
Schopf are prokaryotes. Prokaryotes are the type of cell that all
bacteria are. Unlike eukaryotes, prokaryotes have no nucleus.
The DNA chain lies free as a loop in the cytoplasm of the cell.

It was not until about 2.8 billion years ago that eukaryotic cells
appeared on the earth. These are the nucleated cells of which you
and I are made. It was much later still, around 0.6 billion (600
million) years ago that we find the first fossils of multicellular animals. See figure 2.1 for a time line to show more clearly the various dates of importance.

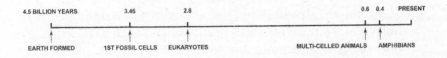

figure 2.1

The most complex system in the universe

It is not an exaggeration to say that cells, including bacterial cells,
are the most complex known systems in the universe. It is all too
easy to dismiss bacteria as simple and just primitive.

Paul Davies, in his book about the origin of life, *The Fifth
Miracle*,[2] has written:

The living cell is the most complex system known to man. Its host of specialised molecules, many found nowhere else but within living material, are themselves enormously complex. They execute a dance of exquisite fidelity, orchestrated with breathtaking precision. Vastly more complicated than the most complicated ballet, the dance of life encompasses countless molecular performers in synergetic coordination.

Should you ever think of reading a textbook on molecular and cellular biology then you will have the fact of the complexity of bacteria confirmed. In the textbook by Stephen L. Wolfe[3] we read:

> The apparent simplicity of prokaryotic cells is deceptive. Most bacteria and cyanobacteria[4] contain complex molecular systems, can use a wide variety of substances as energy sources, and are able to synthesise all their required organic molecules from simple starting substances, such as water, carbon dioxide, and inorganic sources of nitrogen, phosphorus, and sulphur. In many respects, in fact, prokaryotes are more versatile in their biochemical activities than eukaryotes.

Even the plasma membrane that surrounds a bacteria is a very complex piece of apparatus and it carries out numerous functions. The most important of these is transport – water-soluble substances are moved in and out of the cell in a very controlled way. The membrane itself is made from a double layer of lipid molecules but also contains 50 or more different proteins which span the membrane from one side to another.

The transport of ions across the bacterial cell membrane is accomplished by specific proteins which interact with an ion molecule on the outside, bind it at the surface of the membrane, the protein changing shape after the use of specialised enzymes and then somehow 'squeezing' the ion molecule through the membrane and into the cytoplasm.

The membrane also contains proteins that can act as receptors by recognising and binding specific molecules that penetrate through the membrane from the surrounding medium. Binding these external molecules triggers internal reactions that allow the bacteria to recognise their environment.

Those ancient bacteria found in the Apex Chert probably carried out photosynthesis. Photosynthesis involves complex tailor-made molecular systems, harnessing the energy from the sun's photons.

Far too often I hear experts confidently postulate about the first cell membranes. They know that the first life had to have a protective coat to hold the DNA and all the incredibly sophisticated molecular machinery inside the bacteria. Some, including Schopf (who discovered the Apex Chert bacteria) suggest that the first, fledgling and replicating DNA was surrounded by a type of soap bubble. The membrane of a bacterium is however entirely different from a simple soap bubble in structure and these first cyanobacteria had the same dynamically complex cell membranes as modern cells do.

As I will discuss later, the actual details of the membrane must be coded for in the DNA of the cell. It is a major problem for theorists of early life to imagine how the first DNA achieved such a code without first having the membrane around it. Such 'chicken and egg' scenarios bedevil the understanding of early life.

Within the cell is the cytoplasm which contains large numbers of ribosomes, which are small spherical particles. These contain more than 50 different proteins in combination with several types of ribonucleic acid (RNA). They are the sites where amino acids are assembled into proteins.

Cyanobacteria also have bag-like sacs in the cytoplasm known as gas vacuoles, evidently used in buoyancy, holding the cells near liquid surfaces or sources of light. The cytoplasm also contains deposits of lipids, polysaccharides and inorganic phosphates. These phosphates may act as energy reserves or as raw materials for nucleic acid or phospholipid synthesis.

My reason for detailing these facts about the humble bacterium (before we even begin to look at its DNA) is to emphasise how astonishingly intricate these first cells are. Those fossils of 3.46 billion years ago are just such cells. Let us keep this in mind when we attempt to see how they might have come about by chance. Current evolutionary theory demands that they came about by a series of chance accidents without any design whatsoever. We will keep this in mind when pondering the fact that they are the most complex structures in the universe.

Theories of the start of life

Darwin himself had no idea how life might have started on the earth. It was a mystery to him until his death. He certainly pondered over the problem and wrote in 1871 to his botanist friend Joseph Hooker:

> It is often said that all of the conditions for the first production of a living organism are now present, which could ever have been present. But if (and oh! what a big if!) we could conceive in some warm pond, with all sorts of ammonia and phosphate salts, lights, heat, electricity etc. present, that a protein compound was chemically formed ready to undergo still more complex changes, at the present day such matter would be instantly devoured or absorbed, which would not have been the case before living creatures were formed.[5]

This was written 12 years after the publication of his *Origin of*

Species. I hope that I can show that his optimism about a warm pond accidentally producing life from basic chemicals was totally unfounded.

A great deal of research has gone on to try to simulate the conditions of the early earth and to see how the first building blocks of life might have formed. It has been found that organic compounds and carbon are common in the solar system and beyond. Indeed some meteors, such as the famous Murchison meteorite, contain hydrocarbons, urea, ketones, alcohols, amines and amino acids. Many therefore believe that the organic compounds necessary for life came from space. Recently, techniques using radio spectral lines have revealed many organic molecules in interstellar space. And so, the basic organic molecules, from which life is made, would therefore appear to have been present on the early earth.

I marvel however at how some scientists view these facts, about the chemical ingredients of life being present as if the whole problem is almost dealt with. It is true that the ingredients existed. It is also true that the ingredients to make a Roll Royce car can be found lying in a scrap heap. I do not then jump to the conclusion that a Rolls Royce will somehow appear out of the heap without a fair amount of help from an engineer.

In 1953 a very famous experiment was performed by Stanley Miller, in collaboration with Harold Urey, at the University of Chicago. I well remember that the details of this experiment were in most of the biology textbooks which I used in the past. They claimed to have built the building blocks for life in a laboratory. As such, it is still referred to by many as another piece of evidence for evolution.

What they did was to reproduce what they thought was the atmosphere of the very early earth and put this mixture of hydrogen, methane and ammonia in a sealed glass container. Water was added which was boiled. They also discharged electricity through it with spark plugs to provide energy.

After one week they analysed the sediment and found many organic molecules including 10 of the 20 amino acids that occur in proteins, fatty acids, aldehydes, sugars, purines such as adenine and guanine as well as pyrimidines uracil, thyamine and cytosine. These purines and pyrimidines are important constituents of DNA.

These results were understandably sensational and have been used endlessly to argue for the entirely natural, material processes of forming life.

However, it turns out that the atmosphere they used was almost certainly incorrect – it is now generally accepted that the early atmosphere was a mixture of carbon dioxide, nitrogen and water. These were the products of 'outgassing' from volcanoes.

When the experiment is repeated with the correct atmosphere then only the amino acid glycine is produced and and none of the building blocks of DNA. Many people do not realise this. When the correct atmosphere is used in the experiment then we do see some organic material forming. Small amounts of hydrogen cyanide (HCN) and formaldehyde (HCHO) are formed and some scientists such as Bernal[6] propose that these compounds may have been concentrated on the surface of clay particles. Experiments using UV irradiation and electrical discharges do produce some amino acids when these conditions are met.

Where does this leave us? I have detailed these experiments to show you how far scientists have got with the whole issue. The fact is that they have barely begun to find any solution. Even if amino acids are formed and available, this does not mean that proteins can form from them. The spontaneous formation of amino acid in a chemical mixture is a 'downhill' process - it does not require extra energy to accomplish it. However, coupling amino acids together to form peptides (and then proteins) is very much an 'uphill' process. Each peptide bond that is made requires a water molecule to be removed from the chain. This is thermodynamically unfavourable and it will not happen spontaneously in a watery medium like the primordial soup.

In fact, as we will see, proteins are made quite differently in living organisms. They are coded for in DNA and the bonding of amino acids is only possible because of customised molecules at the site of ribosomes in the cell.

DNA

Let's have a look at that incredible molecule known to us all as DNA (deoxyribonucleic acid). The structure of DNA is the double helix discovered by Crick and Watson in 1953 – 50 years ago as I write. It consists of two strands of molecules coiled into a helix with cross-links holding them together called nucleotide bases. These bases come in four different types known as adenine, guanine, cytosine and thyamine – or AGC and T for short. Each link between the two strands actually consists of a pair of bases (one from each strand) joined together in a chemical bond.

A can only fit with T and G can only fit with C. This means that any strand of DNA on its own is a template for a complementary strand that fits with the correct bases. For example, a sequence such as GTACCATG must link exactly with a complementary sequence consisting of CATGGTAC.

And so, that is the basis for the way in which DNA is replicated during cell division. When the helix is unzipped, exposing the unlinked bases, then free molecules of AGC and T will join up to form a new chain. This replication process works very well and uses a number of specialised enzymes to facilitate the operation.

The amount of genetic information in the DNA is amazingly large. A bacterium such as *E. coli* has a few million pieces of information in its DNA. The entire collection of information is known as the genome.

Building proteins

Proteins are the main building blocks of life and are quite different from DNA. The DNA in a cell contains the nucleotide sequence that provides the information to build proteins.

Proteins are chains of amino acids of which there are 20 in all. Each of the 20 amino acids is coded for on the DNA. Although there are four different bases (A,T,G and C) the amino acids are coded for in triplets known as codons. A typical protein may have thousands of bases on the DNA representing its chain of amino acids.

To turn this information into proteins requires the help of a closely related molecule known as RNA (ribonucleic acid). This is also made up of four different bases but instead of T (thyamine) there is U (uracil), which serves the same purpose.

There are various forms of RNA but the one which reads the recipe for a protein is messenger RNA (mRNA). The mRNA molecule is formed by linking up with the complementary bases from the DNA sequence and is therefore a copy of that sequence.

The mRNA then leaves the DNA and goes to the ribosomes which are the factories for assembling proteins. The mRNA is read in the ribosome like a tape recording and at each triplet of bases another amino acid is linked on until the correct chain is formed to make a protein.

The amino acids are brought to the ribosome attached to another form of RNA called transfer RNA (tRNA). Each particular tRNA brings only one type of amino acid to the production line. The tRNA exposes its triplet of bases and this links up with the complementary bases on the mRNA, thus constructing the molecule of protein. When the process is complete the ribosome receives a 'stop' message from a triplet of bases on the mRNA that instructs the ribosome to cease further construction.

I have outlined only the bare basics of what is a marvellous and intricate system of coding and production.

Building Nucleic Acids

Many scientists now feel that the first form of replicating nucleic acid was RNA and not DNA. They envisage a sort of early 'RNA world'. This is because DNA itself is synthesised from chains of RNA. Also RNA can have catalytic activity similar to protein enzymes and is used, as described above, in the translation of genetic information into the formation of proteins at the ribosomes.

The big question at this point, as we look at these complexities, is how could the DNA and RNA arise *de novo*?

Now we have seen that some of the building blocks are produced from HCN (hydrogen cyanide) in a manner that could have occurred naturally. In particular we can imagine the purines forming, adenine (A) and guanine (G), but not the pyrimidines (T,C and U).

Many feel that this is tantamount to solving the problem but they are wrong. The building of nucleic acids (DNA and RNA) from these units requires very, very unlikely steps which have never been achieved experimentally and which are even theoretically virtually impossible to envisage. One solution that has been proposed is that a nucleic acid that is easier to construct was a precursor of RNA and DNA – such as threo-nucleic acid (TNA). However there is no real evidence that any pre-biotic soup could produce TNA either.[7]

The steps needed for making nucleic acids require the formation of nucleosides (purine or pyrimidine base plus the 5-carbon sugar ribose), then the phosphorylation of the nucleosides to give nucleotides and then the polymerisation of these units to form nucleic acids capable of self-replication.

Joyce[8] and many others have concluded that the problems are insuperable and that life did not start with nucleic acids. In his review in the journal *Nature*, Joyce carefully describes the unlikely steps needed to form a single nucleotide. He also showed that when such mononucleotides were made they would be both right- and left-handed (D and L conformation). When this happens, the nucleotides of different handedness bind and interfere with each other so that the formation of a chain of such nucleotides is impossible.

Because of these difficulties many are looking for some simpler replicating system which would then have introduced the order needed to process nucleic acids. This does seem to me, however, to be simply shifting the problem elsewhere and not solving it.

Tracing the tree of life

Carl Woese, a leading scientist in this field, wrote a paper in 1998[9] about his attempts to trace the 'tree of life' back to its roots in order to find a 'universal ancestor'. A central part of evolutionary theory is that there must be an organism which is the ancestor of all and from which the tree of life has grown. By looking at the molecular sequences of RNA in different organisms and knowing the mutation rates, he has tried to trace the origins of the three main branches of cell life. These are the bacteria, archaea (a group of micro-organisms that live in extreme environments) and eukarya (the form of cells that higher life is made of). The technique he used is similar to the way that linguists trace the origins of languages by looking at their differences.

Woese's findings are that there is no clear original root. The tree is unrooted. Moreover, when a large number of metabolic genes are analysed they indicate that any universal ancestor seems to have had a full complement of these genes – meaning that it was already a very complex cell with all the advanced biochemistry (and even flagella) that we know in modern organisms. I was fascinated when I read his paper because it seems to confirm the fossil evidence of very complex bacteria from the start of life.

To try to overcome this scenario (which, of course, is quite unacceptable to modern scientific materialism), Woese proposes a 'genetic annealing model'. In this model (for which there is no evidence) he proposes that there may have been a pre-cell world when rapidly evolving genetic material was in a flux of high activity with lateral gene transfer (genes moving across from one organism to another). This scenario of his is an attempt to explain the apparent rootlessness of the tree of life and the apparent complexity of the first cells.

My main impression of this idea is its illogicality. To achieve the complexity of a cell then you must have at least a cell membrane to surround and protect the multitude of molecular machines within. Any pre-cell world must have been naked RNA or DNA – which could not survive unprotected and also could not 'evolve' complex cellular features without being a cell already. What role (for example) would a ribosome have unless it was within a cell? Some would suggest that a complex system such as the ribosome had a different role at first and that this is a case of 'pre-adaptation'. However, this does beg the question of how such a complex and beautifully honed system of molecular machinery could have had a role outside of a cell. This is one of countless examples of 'irreducible complexity' in life – the concept is amplified in Chapter 8.

I have already explained the complexity of the cell wall. Any old wall will not do – all cells, for example, must have a complex transfer system in the cell wall in order to maintain life within. To the evolutionist there seems no answer to this.

Life from space?

Because of the seemingly insuperable problems concerning the origin of cells, some scientists propose the idea that life has come intact from outer space.[10]

Such a view though, seems to me to merely put off the difficulties of how life in outer space may have started. To date there is no evidence for life apart from on earth in our solar system, nor anywhere else in the universe. In fact, the conditions for life that we find on earth, as already described in the preceding chapter, seem to be very, very fortuitous and unlikely to be reproduced elsewhere.

Gene copying processes

Another major problem in evolutionary theory of early life is to do with the errors that occur when the genome is copying itself during replication. In all modern organisms, sophisticated proofreading and error correction mechanisms are used to keep the error rate very low. There are a host of enzymes (which are proteins) involved in this copying process.

According to evolutionary theory, such enzymes did not exist in the first organisms because they had to evolve over time. This would have made the copying very error prone and, to avoid catastrophe, the first replicating molecules would have to be very short. But if it is short then it cannot store enough information to build the copying information itself. As Paul Davies puts it in his book *The Fifth Miracle*:

> A short nucleic acid sequence, as envisaged by evolutionists, would have no chance of containing the information needed to code for the copying enzymes that it needs. Complex genomes require reliable copying, and reliable copying requires complex genomes. Which came first?[11]

This is another of the chicken and egg problems that surround evolutionary theory. The other we have mentioned is the question of DNA needing a cell membrane to survive and the membrane needing DNA to manufacture it.

Chapter Two

The message in the molecules

How does meaningful information get into the DNA? This is a profound difficulty. All life has DNA which is crammed with meaningful information.

Computer scientists describe two sorts of information; syntactic and semantic. Syntactic information is raw data with no meaning. A good example is a snowflake (or any crystalline structure). A snowflake is a complicated arrangement of hexagonal ice crystals. There is information there but it is syntactic – meaningless. Yes, it is obviously reasonably complex and ordered in hexagonal shapes, but all without any message.

In the same way, when you hear a hissing noise of interference on your phone line, you are hearing a form of sound; which is information, but it is without any meaning. Like the snowflake it is syntactic.When you hear someone speaking on that phone however, you are hearing semantic information which is full of meaning.

In the same way, the information that you put into your computer with software is meaningful and therefore semantic. The computer will not run without meaningful instructions which have been designed by an intelligence outside the computer.

The information in the bacterial genome is semantic or meaningful. How did it get there?

Somehow, says the evolutionist, the meaning in the DNA came about by chance, gradually. But I will doggedly pursue this question – where from? The evolutionist must reply: from the natural environment. But this is merely a way of off-loading the problem. How did such amazing meaning come from the environment?

I once again, come to the conclusion that an intelligent designer was at work. There is no other explanation.

The selfish gene

Some evolutionists seem to almost invest the DNA molecule with personality. Richard Dawkins writes about the 'selfish gene'.[12] The entire meaning of life, according to the more reductionist evolutionists such as Dawkins[13], is bound up in the fact that genes (sequences of nucleic acids in the DNA), have an inbuilt propensity or drive to replicate, at all costs. Presumably, if we follow his arguments, this inner urge to replicate at all costs, must have been why the fledgling chains of DNA or RNA determinedly organised themselves to do so – even as far as inventing cells around them.

There is an enormous absurdity about this line of argument. Why should inanimate molecules have any propensity or drive to replicate? Why should they have such a drive that they end up forming impossibly complex structures around themselves? Where does such a drive originate? There is no answer to this. It is true that crystals have a form of replication – repeating themselves when the conditions are right. Crystals, however, have no ability or propensity for manufacturing cell walls, proof-reading enzymes and information-rich meaningful codes. I believe that the idea of molecules having any particular drive is nonsense and needs to be challenged. Yet, the whole of modern Darwinian theory depends upon it.

God of the gaps?

I am well aware of arguments that many will line up against this reasoning. One of the principal ones is the 'God of the gaps' idea. In this argument it is stated that when there are gaps in our knowledge then Christians and others will simply put God in there and say he did it – which is fine until further knowledge reduces the gaps and God is gradually eased out altogether.

I get quite annoyed at this oft-repeated accusation because it is in fact intellectually rather lazy. It is often put out by Christian scientists who believe in 'theistic evolution'. They accept all of Darwinism but say that God sustains and works through the natural laws that drive evolution. No Christian could disagree that he sustains all of nature but to say that he has left no evidence whatsoever of his involvement in the process (as they insist) does not seem to worry them or seem contradictory. Are they saying that God deliberately made evolution appear accidental when it is not? They rule out the supernatural in biology and are wedded, as much as any atheist, to an absolutely, and exclusively, natural mechanism for the appearance of life. As such they are very critical of the growing 'intelligent design' movement that is arising from increasing numbers of academic scientists and philosophers. It would be dangerous to label the intelligent design movement purely as a cranky movement from the United States.

Our knowledge about molecular biology and the first organisms is growing all the time – and is vastly greater than when Darwin lived. This greater knowledge has done the very opposite of easing God out. The more we know and understand, the greater are the mysteries and unexplained facts. The gaps do not, in fact, go away but become more mysterious and the problems seem insoluble on a purely random materialistic basis.

Summary

- At the earliest possible time on the earth we find fully formed complex cells.
- The first RNA or DNA needed a membrane to surround it as a cell. Yet all membranes are encoded in the DNA. Neither can exist without the other – one of the 'chicken and egg' problems with evolutionary theory.
- Biochemists have concluded that RNA or DNA arising

randomly from scratch is so unlikely that it seems theoretically impossible.

● Making functioning proteins from scratch is no more likely to have happened and requires functioning RNA or DNA to be accomplished anyway.

● The 'tree of life' has been found to be rootless, with molecular evidence of fully complex systems in place from the very beginning.

● The copying processes of genes require sophisticated proof-reading enzymes to avoid errors and catastrophe for the DNA. Such enzymes must be coded for in the genes (DNA). One cannot exist without the other – another 'chicken and egg' problem.

● The semantic information in DNA must come from somewhere. Evolutionists have to point to the environment. However, there is no logical way the environment can provide such deep meaning without design.

● Evolutionary theory requires the idea of an innate drive within DNA to reproduce and make complex structures around it. There is no evidence that any such drive exists in a purely materialist molecular world.

And so as I investigated the evidence for the origin of life I found that science actually has no answers to how it all began. It would seem that we have not got much further than the vague ideas that Darwin had with his 'warm pond'.

I trust that some of the material I have discussed in this chapter will help you as a reader to appreciate just what a mystery the start of life is.

1. J. William Schopf (2001). *Cradle of Life: The discovery of Earth's earliest fossils.* Princeton: Princeton University Press.
2. Paul Davies (1998). *The Fifth Miracle.* Allen Lane: Penguin.
3. Stephen Wolfe (1993). *Molecular and Cellular Biology.* Belmont, CA: Wadsworth Publishing Co.

4. One of the types of bacteria in the Apex Chert.

5. Quote from: Michael Denton 1986. *Evolution: A theory in crisis. Adlere& Adler.* P.249.

6. J. D. Bernal (1967). *The Origin of Life.* Cleveland, OH: World Publishing Co.

7. See Simon Conway Morris (2003). *Life's Solution. Inevitable Humans in a Lonely Universe.* Cambridge University. Press. Conway Morris provides an excellent review of the work done in looking at the origin of the first cells.

8. G. F. Joyce (1989). 'RNA evolution and the origins of life', *Nature* 338,pp.217-223.

9. Carl Woesse (1998). 'Proceedings of the National Academy of Sciences.' USA, 95, 6854.

10. See C. Wickramasinghe (2001). *Cosmic Dragons. Life and Death on our planet.* Souvenir Press.

11. Davies, *The Fifth Miracle,* Allen Lane: Penguin. P.32

12. Richard Dawkins (1989). *The Selfish Gene.* New edition. Oxford Univ. Press.

13. Richard Dawkins (1995). *River Out of Eden.* Weidenfeld & Nicolson, London

Chapter Three

An Explosion of Life

Charles Darwin admitted that one of the great unresolved problems of his theory was the fact that there was no fossil evidence of animals before a time known to him as the *lowest Silurian* but which to us is called the Cambrian period. He knew that in the Cambrian rock strata were many different forms of animals but that there were none at all seen before. He said, in *The Origin Of Species*:

> If my theory be true, it is indisputable that before the lowest Silurian stratum was deposited, long periods elapsed, as long as, or probably far longer than the whole interval from the Silurian age to the present day; and that during these vast, yet quite unknown periods of time, the world swarmed with living creatures.[1]

He believed, quite rightly, that if his theory was to stand, the fossil record would one day reveal the host of animals that were precursors to the rich Cambrian fauna. His theory rests on the fact that organisms have evolved gradually over long stretches of time and that to achieve such complex creatures as those of the Cambrian period would require a clear, earlier, succession of slowly changing fossils. Darwin therefore, very honestly, put forward a test of his theory, which if proved negative would be fatal to it. This, as the philosopher Karl Popper has explained,[2] is the true mark of a scientific theory. A theory must be potentially fal-

29

sifiable. There must be clear tests to a theory, which, if failed, would reject that theory. One such test is that proposed by Darwin – that future discoveries would be bound to uncover the myriad of beasts that led gradually to the Cambrian fauna. In this chapter I describe the present-day knowledge of this fauna and what we know about any precursors to it in the strata.

We now know that the rocks of the Cambrian were deposited on the Earth from 543 million years ago. See figure 3.1 for the geological timescale.

Eon	Era	Period	Epoch	Date at beginning (Myr)
Phanerozoic Eon				
	Cenozoic Era			
		Quaternary Period		
			Holocene Epoch	0.01
			Pleistocene Epoch	1.6
		Tertiary Period		
			Pliocene Epoch	5
			Miocene Epoch	23
			Oligocene Epoch	35
			Eocene Epoch	56
			Paleocene Epoch	65
	Mesozoic Era			
		Cretaceous Period		146
		Jurassic Period		208
		Triassic Period		250
	Palaeozoic Era			
		Permian Period		290
		Carboniferous Period		362
		Devonian Period		408
		Silurian Period		439
		Ordovician Period		510
		Cambrian Period		550
Precambrian				4560

figure 3.1

About 85 per cent of the Earth's history had already passed by the time the Cambrian rocks were laid down. As we have seen, the first organisms were bacteria, appearing about 3.8 billion years ago. These non-nucleated, asexually dividing bacteria, the prokaryotes, were the only creatures around until about 2 billion years ago when the eukaryotic microbes appear as chemical fossils.

Eukaryotic cells are much larger than prokaryotes, have their DNA in nuclei, they have organelles such as mitochondria within the cell (or chloroplasts for photosynthesis in plant cells) and they can reproduce sexually. All animal, plant, fungal and algae cells are eukaryotic. There are a variety of theories as to how the eukaryotic cell arose but it is not my purpose here to go into the details. Suffice it to say that there are enormous differences between prokaryotes and eukaryotes – not least the means of cell division – and there is no clear understanding of how such cells might have evolved –though the most popular theory at present is that mitochondria (and chloroplasts) are remnants of bacteria which entered the cell in some form of symbiotic relationship. It would be about another 1.5 billion years before any multicellular animals came onto the scene.

In 1909 Charles Doolittle Walcott, secretary of the Smithsonian Institution, was looking for fossils in a remote area of British Columbia and came across an amazing collection[3]. They were contained within a formation known as the Burgess Shale and they were dated as being from the mid-Cambrian period, 520 million years ago. These fossils, he soon realised, were very unusual because the preservation was superb and they included many soft-body structures of animals. Under most conditions a soft-bodied animal fails to be fossilised but the conditions had been special here – probably these sea creatures had been buried rapidly in mud slides and preserved in anoxic conditions. Most fossil collections contain only the hard parts of organisms (external or internal skeleton) and therefore only a small percentage of animals are preserved. Here, however, in the Burgess Shale, was a very complete collection which showed the range of the fauna existing then. It was a magnificent window into the Cambrian, sea- bottom dwelling creatures.

Walcott collected furiously, year after year, and he came across a host of animals that were very hard to classify. Walcott classified animals according to known major groups, known as phyla. A phylum has a unique type of body plan and all existing animals fall into about 35 of them. Our own phylum, the chordates, is as distinctive as the rest.

We now know that the Burgess animals include a large number of body plans (phyla) that are unique and that are not represented at all in modern animals (though there is dispute about this as I shall explain). Walcott however managed to 'shoe-horn' all his finds into existing phyla. It is believed that he did this because to do otherwise was to flout the existing understanding of evolution. According to Darwin, the tree of life had one or two roots and gradually diversified over time to produce many branches. It had the shape of an inverted cone (point downwards). For anyone to suggest that at the very beginning of multicellular life there were more phyla than at present, would have been a form of heresy. It would mean that the cone (or the tree) would have to be turned upside down. The very processes of evolution, according to Darwinian theory, require gradual diversification. However, the Burgess Shale animals reveal the very opposite, as I shall describe here.

That a leading scientist such as Walcott should bend the facts in such a way reveals something very important about scientists. Scientists are not the utterly objective creatures that we like to imagine. Like everyone else, they have many preconceived ideas and agendas which often colour their interpretation of the data. There is no guarantee at any time that the leading scientists of any field of study are seeing things objectively – no more today than back in 1909. It is too easy to dismiss scientists of a bygone era (such as Walcott) as somehow less rigorous than ourselves. No doubt scientists of his day were thinking the same of their forebears. Let this be a warning – particularly when there is any body of opinion which becomes almost sacred, a sacred cow indeed. Evolutionary theory is decidedly one of these areas.

It was only in the 1970s that a group of British and Irish palaeontologists began to look again in depth at the range of specimens from the Burgess Shale. Whittington, along with Briggs and Conway Morris,[4] began to examine systematically the thousands of specimens that were then available. What they discovered was stunning and sent shock waves throughout the palaeontological community. They began to describe, with painstaking care and professionalism, animals that were bizarre and quite alien in appearance. Nothing like them had ever been seen or imagined before. Alongside more familiar body plans were some that were more like hallucinations than specimens – hence the wonderful name of *Hallucigenia* for one beast which had seven pairs of spines on one side, seven tentacles on the other and was so novel in shape that even now we do not know for sure which is the front or the back. There was *Opabinia* with five eyes and a frontal 'nozzle'. There was *Wiwaxia*, a flattened, oval creature of great beauty, about three inches long, its body covered in plates and spines. Many others of these small sea creatures were of totally unclassified body plans. All are now extinct.

All in all there are between 15 and 20 sea animals in the Burgess Shale that have anatomies quite unlike any known phyla and that probably represent entirely separate phyla. What is more, almost all our modern phyla are also represented in the Burgess fauna[5] and those that are not appear relatively soon after in the geological strata. Even our own phylum, the chordates, is seen – represented by the Burgess Shale species called *Pikaia*. It has a notochord (equivalent to our vertebral column) and the zigzag muscle pattern typical of the chordates.

We see an early appearance of the trilobite – an arthropod

which became ubiquitous for long afterwards, becoming extinct about 250 million years ago. It is worth spending some time describing the trilobite in order to see what a very remarkable animal it was (see figure 3.2).

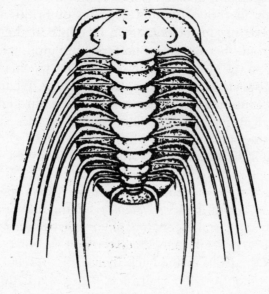

figure 3.2

As I write I have a trilobite fossil beside me, a beautiful creature just over three inches in length and looking superficially like an enormous wood louse. I will give here a basic description of this magnificent animal – which is needed for us to appreciate the absolute mystery as to its sudden appearance.[6]

Roughly oval in shape, wider at the head end, the trilobite had a hard external cuticle made of calcite. There was a head, thorax and tail made from segments of the cuticle. The trilobite takes its name from the unique three-lobed, longitudinal divisions of the body. Limbs were attached to the lower surface. The cuticle is in two main layers of microcrystalline calcite with ability to resist tensile and compressive forces and the whole body was able to flex or roll up in a defensive attitude. The cuticle had numerous pores or canals leading from inside to the surface. These were probably mainly sensory, carrying small hairs externally, each connected to the central nervous system by a nerve running up the canal. Special ridges associated with some of the canals are thought by many to represent a current monitoring system, sensitive to change in water direction. The head is graced with two antennae and there is a mouth leading to a stomach and gut. Internally, scars on the cuticle show where muscles attached to enable the trilobite to move and flex. However, it is perhaps the eyes of the trilobite which excite our admiration most. They are the most ancient visual system known but are as intricate as any compound eye of a modern insect. These compound eyes, one on each side of the head, were composed of multiple radially arranged visual units pointing in different directions and encompassing a wide-angled visual field. The internal structure of the eyes is not preserved but can be assumed to be similar to that of other arthropods. In other arthropods each unit consists of a cylinder of cells capped by a lens (as in trilobites) and with another optical apparatus just below that known as the crystalline cone. These focus light on the rhabdom, which is the photosensitive area at the base of the cylinder. Here, photosensitive pigments trigger electrical discharges to nerves which travel to an optic ganglion. At the ganglion some form of integrated image is produced from the mosaic effect of light from the multiple lenses. In trilobite fossils what is preserved is the external lens of each unit, which is in contact at the edges with the other lenses and is covered by a corneal membrane. The lens is thin and biconvex and is a single crystal of calcite. All the lenses are carefully orientated at the visual surface so that light is not broken into two rays but continues unaltered to the visual pigments within. Different species of trilobite had some differences in the arrangement of these eye elements but all were according to the basic plan.

I have only touched on some of the details we know about the trilobite. It is one of many entirely different but equally complex body plans appearing in the Burgess shale. For such profoundly beautiful and intricate animals just to appear in the strata, in a geological instant, is not just a problem for evolutionary theory but is virtually a demolition of it.

Every species we know is classified (a process know as taxonomy). The system used was established by Carolus Linnaeus in 1758. Every distinguishable form is given a genus (plural genera) name and a species name, such as *Homo sapiens, Tyrannosaurus rex* or *Canis familiaris.* The generic name comes first, followed by the species name. A living species is generally defined according to the biological species concept, as all the members of different populations that naturally interbreed, and produce viable (i.e. fertile) offspring. Fossil species are usually defined according to their anatomy (morphology) – the *morphological species concept* – as assemblages of forms of similar geological age that have very similar skeletons, and which differ from all others by clear characters. Species are grouped together in genera, and each genus may contain one or more species. Genera are then grouped in families, families in orders, and so on. The classification of humans gives us an example:

Species	*sapiens*
Genus	*Homo*
Family	Hominidae
Order	Primates
Class	Vertebrata
Subphylum	Chordata
'Superphylum'	Deuterostomia
Kingdom	Animalia

As mentioned before, the Burgess animals have representatives of our modern phyla, including the chordates, but what is astonishing is that the range of phyla is apparently far greater in the Burgess fauna than in our modern animals. (As I explain later, there is controversy about this idea but I base this argument on the work of Stephen Gould in his book *Wonderful Life*). For example, in the arthropod superphylum there are specimens from each of the four major known arthropod phyla: trilobites (now extinct), crustaceans (crabs, lobsters, shrimps), chelicerates (includes spiders and scorpions) and uniramians (includes insects). But, and this is the amazing fact, there are 20–30 other kinds of arthropod body plans in the Burgess group which are not placed in any modern group at all. Taxonomists have described about 1 million species of arthropods – but all within the four known phyla. The Burgess shale has more than 20 completely different arthropod phyla to add to these four.

Not every palaeontologist agrees with Gould. Some have tried again to somehow harmonise the variety of phyla into some sort of lineage, maintaining that the disparate groups are merely stages on the way to our own extant phyla. Simon Conway Morris is a case in point – I outline the main points of his arguments at the end of this chapter.

Despite the greater variety and number of basic body plans, within each phylum of the Burgess fauna there are actually relatively few species. Diversity of species is low but diversity of body plans (or phyla) is high. This diversity of basic body plans is known in taxonomic circles as disparity. Modern animals are restricted to fewer phyla but have many more species. The big question is how can one explain such an enormous disparity of body plans so early in the 'evolution' of life. It seems to turn Darwin's evolutionary tree literally on its head. Remember that in evolutionary theory there ought to be less range of body plans at the beginning and gradual diversification with time. The very opposite is found in the Burgess Shale.

The Burgess animals are not some localised freak but have been found also around the world in the Cambrian strata of

Greenland and China – as well as in sites closer to the Burgess Shale. What were the precursors, if any, to these Cambrian sea animals? The major candidates as precursors are what are known as the Ediacaran fauna.

The Ediacaran fauna were discovered in 1947 in the sandstones of the Ediacara hills of South Australia. They are dated as just pre-Cambrian, from 575–550 million years ago. These are the earliest known multicellular animals and they have now been seen in areas all over the world. The majority of forms are soft-bodied and are of a variety of shapes. None have the complexity of the later Cambrian fauna. What is troubling to the taxonomist is that they do not seem to have any clear relationship to the fauna of the Cambrian explosion. They appear consistently in the strata of the same age around the world and then seem to have become extinct. The fauna of the Cambrian explosion appear just after these Ediacaran creatures.

We have then the first multicellular animals of the Ediacaran type, their presumed extinction (though most experts believe the Ediacaran fauna were precursors of the Cambrian) and then the sudden appearance (sudden by palaeontological standards) of the Cambrian animals – all within about 50 million years. It must be stressed that 50 million years is a very short period in palaeontology, when looking at such major changes in the fossil record.

To counter claims that the Cambrian explosion represents a threat to Darwinian theory, some have maintained that the only reason we see so many forms suddenly in the Cambrian explosion is because it is a special site which shows up soft-bodied animals. They maintain that the animals were there before but had not been preserved because of having soft bodies. There are very strong arguments against this. Firstly, there are plenty of sediments from earlier periods which preserve soft bodies and do not have the animals of the Cambrian explosion – these are the fossils of Ediacaran fauna which are soft bodied animals. Secondly, many of the Cambrian explosion animals have highly developed hard exteriors (for instance, the host of arthropods), which do not appear in the early strata. Thirdly, the animals of the Cambrian explosion do not appear only in one place but have been seen to suddenly appear in sediments around the world – not as beautifully preserved as in the Burgess Shale but nevertheless the same fauna and appearing at the same geological time.

Stephen Jay Gould was Professor of Zoology and Professor of Geology at Harvard University and was a prolific writer on matters to do with evolution. In his book *Wonderful Life*, he wrote concerning Darwin's problem of the sudden appearance of animal life in the Cambrian:

> Darwin invoked his standard argument to resolve this uncomfortable problem: the fossil record is so imperfect that we do not have the evidence for most of life's history. But even Darwin acknowledged that his favourite ploy was wearing a bit thin in this case. His argument could easily account for a missing stage in a single lineage, but could the agencies of imperfection really obliterate absolutely all evidence for positively every creature during most of life's history? Darwin admitted: 'The case at present must remain inexplicable; and may be truly urged as a valid argument against my views here entertained' (*Origin of Species*, 1859).
>
> Darwin has been vindicated by a rich Precambrian record, all discovered in the last 30 years. Yet the peculiar character of this evidence has not matched Darwin's prediction of a continuous rise in complexity toward Cambrian life, and the problem of the Cambrian explosion has remained as stubborn as ever – if not more so, since our confusion now rests on knowledge, rather than ignorance, about the nature of Precambrian life.[7]

Gould tries to have it both ways, saying Darwin has been vindicated but then refuting this by pointing out the fossil record that we do have which shows no Darwinian gradual progression of animals. As already described, we do know of fossils before the explosion but these completely overturn the Darwinian model. For 3.2 billion years we have only microbial life. Then, in a period of about 50 million years: the Ediacaran fauna which appear unrelated to the Cambrian fauna followed by the host of body plans of the Cambrian explosion. Let me quote Gould again from later in the same chapter of his book:

> Thus, instead of Darwin's gradual rise to mounting complexity, the 100 million years from Ediacara to Burgess may have witnessed three radically different faunas – the large pancake-flat soft-bodied Ediacara creatures, the tiny cups and caps of the Tommotian[8], and finally the modern fauna, culminating in the maximal anatomical range of the Burgess. Nearly 2.5 billion years of prokaryotic cells and nothing else– two-thirds of life's history in stasis at the lowest level of recorded complexity. Another 700 million years of the larger and much more intricate eukaryotic cells, but no aggregation to multicellular animal life. Then in the 100-million-year wink of a geological eye, three outstandingly-different faunas – from Ediacara, to Tommotian, to Burgess. Since then, more than 500 million years of wonderful stories, triumphs and tragedies, but not a single new phylum, or basic anatomical design, added to the Burgess complement.
>
> Step way back, blur the details, and you may want to read this sequence as a tale of predictable progress: prokaryotes first, then eukaryotes, then multicellular life. But scrutinise the particulars and the comfortable story collapses. Why did life remain at stage one for two-thirds of its history if complexity offers such benefits? Why did the origin of multicellular life proceed as a short pulse through

three radically different faunas, rather than as a slow and continuous rise of complexity? The history of life is endlessly fascinating, endlessly curious, but scarcely the stuff of our usual thoughts and hopes.[9]

It must be stated that Gould remained a convinced, though rebellious, Darwinist but he is being very honest here about the problems faced by the theory of evolution.

There are seven main explanations given by evolutionary theorists for the sudden explosion of life in the Cambrian period:

1) The evolutionary ancestors of the Cambrian fauna were very small and therefore not seen in the strata. The palaeontologist Richard Fortey, rather lamely, gives this idea in his popular book *Life – An Unauthorised Biography*.[10] The fact is that palaeontologists are very capable at searching for tiny, microscopic fossils and have never found such precursors, despite intense efforts.

2) The Cambrian explosion is the result of there being no competition in the struggle for life because there were no other animals before this time to fight for ecological space on the planet. This resulted in the free expression of many different evolutionary experiments – which survived because there were no competitors. Known as the 'first filling of the ecological barrel' this idea has permeated most comment on the reason for the explosion. Gould and others have, however, criticised this view strongly, and in my opinion correctly. The problem is that the Cambrian explosion is so huge and diverse across the globe. If organisms have the potential to diversify like this then why have we not seen any new phyla[5] appear since Burgess times? There have been many instances of new environments such as islands rising out of the sea where, initially, there has been no competition at all.

Gould and others are forced to believe that the organisms themselves were different in Cambrian times – with, somehow, more potential for diversification. The next two theories are based on this concept.

3) The third idea sees the genome (the full complement of genes of an organism) as somehow more flexible in Cambrian times. A more flexible genetic code might allow much more free expression of different forms. It is postulated that the Cambrian organisms evolved from a single common ancestor not long before the Cambrian period. This ancestor, it is proposed, had a very flexible genome but after diversification the genomes of the organisms became more rigid and unable to express diverse new body plans.

There is no evidence for this idea and it seems contrived. It would seem very unlikely that every single species in the Cambrian was forever 'locked' into a less flexible genome so that thereafter there could not be one new phylum produced during the next 500 million years.

4) The fourth idea suggests that once an organism appears it is limited in what it can do in terms of change. It is tied to an adaptive peak (imagine the ecological landscape to be a series of peaks). The higher an organism is on a peak, the more successful or adapted it is in that environment and the more complex it is. At the start of the Cambrian there were many peaks on which were 'sprinkled' a variety of organisms. Each organism can only specialise even further and go higher on its particular peak. It is very hard to jump to another peak once the landscape is occupied, the organism is tied to its peak and so changes only a little or becomes extinct.

This idea, which actually is very similar to the first one, fails to explain why there have been absolutely no new body plans (phyla) since that time, given that environments such as new islands have repeatedly arisen to provide an empty landscape (and more opportunities for organisms to arise). It also assumes that literally all possible peaks were filled during that one time in the early Cambrian – a very hard thing to believe. There must have been many other potential body plans to try out even amongst the widely diverse Cambrian fauna. Also, given that there have been a series of major catastrophic extinctions of phyla repeatedly over the past 500 million years, surely there was opportunity at these times for more variation in the newly vacated ecological niches.

5) Because what we see in the Cambrian explosion seems to contradict Darwinian evolution, some experts see the variety of body plans as merely steps in the evolution of our existing phyla (as I mentioned above). They see the very strange phyla to be nothing other than precursors of our modern ones. This process, they say, took about 40 million years. The argument relies on a cladistic framework (cladistics is a way of showing a tree of descent based on shared inherited characteristics). This viewpoint depends on most of the Burgess animals being consigned to 'stem' groups, which later come together into the 'crown' groups of more familiar phyla. It is hard to comment on this idea – it is very much cutting-edge discussion in the palaeontological community. It would seem to be another way of trying to squeeze uncomfortable facts into some sort of understandable framework with little, if any, real data to commend it.[11]

6) In a recent book, Andrew Parker has proposed a new theory as to why there was such an explosion of life.[12] He reckons that all is because of the development of the eye. It is vision, he feels, which accounts for the pressure that was needed to evolve many different forms all at once. Predation, using the first vision, required many new ways of evading and surviving and this, he says, is why evolution went so rapidly at this time. I have read his book carefully and found the idea very forced. In particular, we have no idea how the first visual systems themselves arrived suddenly. The first creature we know with vision, the trilobite, complete with

outstandingly complex eyes, appears after the beginning of the Cambrian explosion. It seems to me just another stab in the dark to try to explain the inexplicable.

7) Some feel that a huge change in climate triggered the Cambrian explosion. One idea, which is not widely accepted, is that there was a very cold period known as 'snowball earth' some tens of millions of years before the Cambrian explosion and the melting of this somehow caused the stress needed. How this could do so is not fleshed out or explained. There is also the problem that quite a long period elapsed after 'snowball earth' before the Cambrian explosion.

Darwinian evolutionary theory fails to explain the Cambrian explosion which remains an enigma – even more now than in Darwin's time. At least he had the hope that more fossils would be found to support his theory. We have now searched the sediments and have a very clear picture of the succession of forms over time. The evidence consistently shows a sudden, riotous appearance of diverse body plans. All our modern phyla were there and almost certainly many more. Once this 'creative' extravaganza was over, there would never again be a truly original body plan (phylum) on our earth. The evolutionary 'tree' has been turned on its head and we are obliged to seek an alternative model.

It may infuriate the Darwinian faithful but my impression of Darwin as a scientist and a basically humble man leads me to think that he would now be seeking a totally different solution, if he were alive and seeing the evidence before us. We, who have this evidence, are led to the conclusion that what we see in nature is not always explicable in purely mechanistic or material terms. We are drawn, in fact, to contemplate the transcendent and the mysterious in life.

1. Charles Darwin (1859). *The Origin of Species.* John Murray
2. Karl Popper (1934). Scientific Method. From *Popper selections.* Prineton Univ. Press
3. S.J. Gould (1989). *Wonderful Life. The Burgess Shale and the Nature of History.* New York: Vintage.

4. Simon Conway Morris (1998). *The Crucible of Creation. The burgess Shale and the rise of animals.* Oxford Univ. Press.

5. Helpful is: Janet Moore (2001). *An introduction to the Invertebrates.* Cambridge University Press

6. See ENK Clarkson (1993). *Invertebrate Palaeontology and Evolution.* 3rd edn. Chapman and Hall.

7. Gould, *Wonderful Life,* P.57.

8 The Tommotian fauna is now considered part of the Cambrian explosion.

9. Gould, *Wonderful Life,* PP. 59-66.

10. Richard Fortey (1998). *Life – An Unauthorised Biography.* Flamingo

11. See Simon Conway Morris (2002). 'Body Plans', in *Encylopedia of Evoloution,* vol. 1. Oxford Univ. Press.

12. Andrew Parker (2003). *In the Blink of an Eye: the cause of the most dramatic event in the history of life.* Free Press.

Chapter Four

The Fossil Record of Invertebrates

Perhaps the most important body of evidence that can make evolutionary theory sink or swim is that of the fossil record. We have in fossils direct evidence of previous generations of animals and plants over the entire history of life. These are hard facts, which can be touched and pored over, and, surely, if anything can prove or disprove present evolutionary theory it is these wonderful objects. I have beside me, as I write, as well as the aforementioned trilobite, fossils of an ammonite (180 million years old) and a brachiopod (160 million years old). Each is entirely different but beautiful. Each speaks to me of a bygone era. They are points of contact with specific times in the past – each lived at a definite time and place. What do they tell us about life's history? I will attempt to outline here what we know based on the palaeontological knowledge we possess, which is so much greater than that which Darwin had.

Once the great explosion of multicellular animals came on the scene in the Cambrian, about 530 million years ago, we can thereafter observe fossils in the strata of sedimentary rocks. The record is unbroken – in the sense that there is no time that is without representative fossils to be found somewhere on the Earth. There have been extinctions (the vast majority of all species are extinct) and a pattern of progression over the years. Some extinctions were global and catastrophic – such as at the end of the Permian period, 245 million years ago, and the end of the Cretaceous, 65 million years ago. The latter extinction is widely known because

47

it was the time when the dinosaurs disappeared. What is less well known is the final demise of the ammonite at the same time (along with many other classes of animal). Such extinctions are thought to be due to massive and sudden changes in the environment, such as those produced by impacting asteroids or comets.

Despite such catastrophes, life continued and the record seems one of 'progress'. We begin with the invertebrate sea creatures, then the rise of vertebrates such as fish, then the appearance of amphibians, reptiles, birds, mammals and man - to say nothing about the amazing profusion of insects, fungi and plants that have appeared over the immense tracts of time.

Darwinian theory requires that there be clear evidence of gradual change from one species to another. There ought to be a seamless transition from one type to another in the fossil record because the theory depends totally on the idea that each creature or plant is a modified descendant of previous generations. Small modifications, if advantageous in promoting fitness, will be selected to survive – and thus lead to new species over time. Darwin was acutely aware of the need to find transitional forms in the strata so that his theory would hold.

Darwin, however, admitted that, at his time at least, the evidence for transitional forms was very poor indeed. He devotes an entire chapter of *The Origin of Species* to this (chapter 9, entitled: 'On the imperfection of the geological record'). He writes:

> Why then is not every geological formation and every stratum full of such intermediate links? Geology assuredly does not reveal any such finely graduated organic chain; and this, perhaps, is the most obvious and gravest objection which can be urged against my theory. The explanation lies, as I believe, in the extreme imperfection of the geological record.[1]

Darwin was a very honest man and he constantly faced direct-ly the various challenges to his theory. To read his *Origin of Species* is to see science at its best, searching and painstaking accumula-tion of evidence with careful argument and the taking account of alternative views. It is clear to me that one of the greatest tests to his theory, as he very well knew, was the fossil record. He recog-nised that it did not support him – yet. He did believe, though, that future palaeontologists would confirm it by their finds after he had gone. He also gives cogent reasons in that chapter as to why the record is imperfect. Some of these are, of course, true. In particular, it is wrong to imagine that over millions of years there has been a constant steady accumulation of sediments to preserve organisms. Sedimentation is not constant – conditions change and fluctuate, often organisms live in places with little possibility of fossilisation, sediments themselves undergo great changes and metamorphosis so that fossils are destroyed in the process. It is therefore quite wrong for anyone to expect every lineage of every branch of the fossil record to be preserved. There are also argu-ments to say that intermediate or transitional forms may have been fleeting or peripheral in many populations – so that the pos-sibility of finding them fossilised is remote.

All these arguments are good up to a point. Yes, we can agree that a perfect record is unobtainable. But what we can argue about is whether the imperfect record we now have (vastly greater than Darwin's) continues to hold a veil over the proceedings or whether we now have at least some better evidence to support or refute the theory of Darwin.

Norell and Novacek[2] tested the completeness of the fossil record. They compared geological evidence about the order of appearance of different groups of vertebrates in the rocks with evidence from cladograms (diagrams of relationships based on shared anatomy of different animals). In most cases they found a good match of geological age and the order in cladograms. They found that the fossil record for land vertebrates is as good as the echinoderms, a group we will look at in this chapter, which is believed to have a very good fossil record. Fishes and tetrapods (four-footed animals) have equally good fossil records.[3] As Benton says in his well known textbook *Vertebrate Palaeontology*:

> Another observation perhaps confirms that palaeontological knowledge is not completely inadequate: dramatically unexpected fossils are hardly ever found now. If the known fossil record were very incomplete, many dramatic finds would be made, dinosaur fossils in the Permian or Tertiary, human fossils in the Miocene, fish fossils in the Precambrian. This does not happen. New finds are unexpected[4].

However, if you pick up a modern textbook of palaeontology you will be surprised at how similar is the position now to what it was in Darwin's day. These textbooks are, in the main, written by convinced Darwinists but it is illuminating to have confirmed the very problems that Darwin struggled with. The following is a quote from a well-known and standard work, *Invertebrate Palaeontology and Evolution*, by ENK. Clarkson[5]. In the following section he writes about the origin of higher taxa (taxa are the various divisions that organisms are classified into – as already described):

> New types of structure, and new grades and systems of animal organisation, appear very suddenly in the fossil record, testifying to an initial and very rapid period of evolution. Such changes may be relatively small, defining new groups at lower taxonomic levels. Much larger changes, of great importance, result in the origination of the higher taxa: orders, classes and phyla. Even in the first representatives of these groups, all the systems of the body, and all the anatomical, mechanical, physiological and biochemical elements therein, have to be precisely and harmoniously co-

ordinated from the very beginning. The origin of higher
taxa, in this sense, remains the least understood of palaeontological phenomena. It is sometimes referred to as mega-
evolution though the term is seldom used nowadays, as it
seems to be another form of macroevolution operating at a
higher level, and it is hard to know where to draw the
boundary. When more is known about the functions of the
genome, and how the information carried in the genetic
code results in a fully developed individual, we shall have
more of an insight into this most critical yet most elusive of
all aspects of evolution.

The links between higher and lower taxa are obscure,
and are but poorly represented in the fossil record, as
exemplified by the diversification of life in the early
Cambrian, where transitional or linking forms are absent.
The geological record gives no indication of such relationships, and our knowledge thereof rests entirely upon the
traditional disciplines of comparative anatomy and embryology, supplemented, to some extent, by chemical taxonomy. But what the fossil record does give is many examples
of the 'instantaneous' origin of new structural plans.[5]

Clarkson is not here just writing about the Cambrian explosion,
he is writing about all the subsequent large-scale changes seen in
invertebrates throughout evolutionary history. It is, in fact, an
astonishing statement from the heart of current academic
palaeontology and it emphasises the mystery that surrounds the
origin of major changes in organisms – which generally occur
dramatically and suddenly without transitional forms.

Later I will show that succeeding generations of organisms,
once the major phyla had appeared, do take with them some of
the characteristics of preceeding ones. This continuity shows us
that there is a form of 'evolution' (that loaded term!) but it is not
the smooth type that Darwin insisted on; rather it consists of a

long series of 'saltations' – sudden developments of new form for which we have no explanation.

Micro- and macro-evolution

It is important here to explain the difference between *macro-evolution* and *micro-evolution*. These are now standard terms used in evolutionary theory. Micro-evolution is the appearance of small changes over time resulting in various varieties of a particular organism. Examples abound and include the famous *Darwin's finches* in the Galapagos Islands or the variations in shape, size and colour in any animal or plant. Darwin particularly studied the varieties of pigeons and other domestically bred animals, as well as flowers. Micro-evolution is a fact – even if we could dispute the use of the word evolution in the term. We see micro-evolution in the fossil record. For example, the major phyla seen in the Cambrian explosion undergo significant changes over millions of years, many of these changes being helpful in dating the various strata because each variety is often specific to a particular period of time. Micro-evolution is the most studied part of the theory because it is seen and obvious. I will spend time in Chapter 7 discussing its causes but suffice it to say that, in general, the changes seen in micro-evolution are due not to any new genetic material but to a reshuffling or recombination of the genes of the existing genetic pool. Most of Darwin's discussion about variation in species is based on his work studying micro-evolution. He therefore based much of his research on something which is unable to produce any significantly new structures because micro-evolution does not involve new information in the genome.

Macro-evolution, on the other hand, is the appearance (generally sudden) of completely new body plans or structures in an organism. The phyla themselves (as already explained) appear abruptly and without precursors but within each phylum we see throughout the fossil record sudden leaps of change – often with entirely different organs and structures. This is macro-evolution and it is this that continues to puzzle palaeontologists because there is no clear mechanism as to how it happens. According to Darwin, we ought not always to find such sudden changes – there should be at least *some* evidence where the sedimentary record is good, to show that, the changes actually occur slowly. Current evolutionary theory considers that such macro changes occur because of mutations in the genetic code (occasional mistakes, sometimes supposedly leading to beneficial new adaptations). I will show in Chapter 7 that mutation theory fails to give any such explanation. If you look at a modern diagram of the evolutionary development of animals over time, you will see these discontinuities between the higher taxa quite clearly. (See figure 4.1 showing vertebrates.)

Let us look at some examples, beginning with invertebrates after the Cambrian explosion. All the major phyla appear abruptly and without precursors. We will begin by studying the echinoderms. Of all the invertebrates, there is a fairly continuous record of the echinoderms over time in the fossil record. This is because they have a calcitic skeleton which readily fossilises. The section that follows (in italics) is technical and some will prefer to skim read it or leave it out.[6]

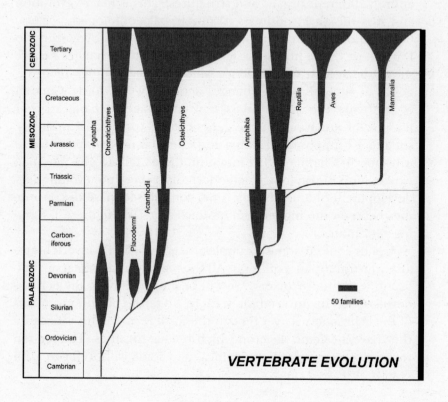

figure 4.1

The echinoderms (see Figure 4.2 at end of chapter)

The echinoderm phylum includes what we know as starfish, brittle stars, sea urchins, sea cucumbers, sea lilies and feather stars. All echinoderms have internal skeletons of calcite plates which are normally spiny and covered outside by skin. Normally, the skeletons have a five-rayed or pentameral symmetry (look at the next starfish you find), though not in all fossil groups, and in some sea urchins a bilateral symmetry is superimposed upon the radial plan. Echinoderms have a unique water-vascular system, which is a complex internal system of tubes and bladders containing fluid. This has extensions which emerge through the skeleton to the outside in the form of tube-feet or podia. These tube-feet have various functions, including locomotion, respiration and feeding. In starfish, for example, sea water enters through a sieve plate and passes by way of a central 'stone canal' to a ring canal and then along radial canals and into the tube-feet. These 'walking' tube-feet of starfish and sea urchins are connected to an 'ampulla', a sea-water reservoir that can be squeezed to extend the tube-feet. The tips of the tube feet stick to the surface on which the animal lies, by suction in sea urchins but by a chemical reaction in starfish. Echinoderms have a nervous system which is a diffuse nerve network but without a brain – although groups of nerve cells can become co-ordinating centres. Perhaps the connective tissue of the echinoderms is the most striking and unique feature. It is capable of rapid and reversible change in stiffness within seconds and is under nervous control. This is due to change in the viscosity of the collagen fibres in the connective tissue, due to ionic movements within the tissue. This results in muscle tone being able to alter without the expenditure of energy, which normal muscle contraction involves. Examples of its use include: the stiffening of sea urchin spines, which can anchor the urchin; the stiffening and relaxation of the body walls of starfish and sea cucumbers; the movement of arms in brittle stars; and the maintenance of tone in the extended arms of sea lilies. There is no blood system, and food and metabolic waste can be transported by the water-vascular system. Gaseous exchange and excretion of metabolites occurs at areas of permeable skin. There is a mouth, gut and anus. There is usually a larval stage – these are ciliated, filter-feeding larvae which are part of the marine plankton.

Such is the basic structure of all the varieties of echinoderm. I mention all this as a reminder of the complexity and the uniqueness of these animals. The various types of echinoderm all appear suddenly in the geological record without any clear transitional forms to tie in relationships. As ENK. Clarkson says: 'The earliest echinoderms were already highly diversified on their first appearance in the Lower Cambrian.'[7] Diagrams of the various echinoderm groups showing their representatives over millions of years show complete separation and no relationship between the groups.

Let us look in detail at just one of the classes – that known as Echinoidea, which includes the common sea urchin. The sea urchin is a 'regular'-shaped member of the class and there is another 'irregular -shaped group, which includes heart urchins and sand dollars. The first examples of the Echinoidea appear in the Ordovician period (510–439 million years ago). They appear suddenly, with no transitional forms known with the earlier Cambrian echinoderms. The basic early forms rejoice in the name Perischoechinoidea and virtually identical urchins exist today (the cidaroids). The sea urchin has an external 'test' comprised of calcitic plates arranged radially around the globular-shaped urchin. Some of these have pores, through which the tubular feet emerge. Spines are attached to the plates. There is a sophisticated ball and joint structure whereby the spines can move on the plates by means of muscles attached to their bases. Amongst the spines there are organs of balance. At the top of the test there is a central hole or 'periproct' through which the anus expels waste. At the bottom there is a central 'peristome' which accommodates the mouth. There are also gills at the region of the peristome. The mouth apparatus consists of a suspended 'lantern' which has five strong jaws, each with a single calcitic tooth. The gonads have an opening near the periproct. The water-vascular system has one opening – beside the periproct. An intricate system of canals leads to the tube-feet. Here there is an ampulla (a sac), just inside the test and from this the canal bifurcates through the test into the foot. The foot is a complex structure consisting of an outside epithelial layer, then a collagen layer and then a muscle layer – all surrounding the water-

filled tube within. At the end is a sucking disc which somehow attaches to the sea floor to aid locomotion. A tube-foot will extend when water pressure within it increases due to contraction of the ampulla. As water comes in, the muscles relax and the foot extends. When the muscle contracts the water is expelled and the foot retracts. The tube-feet are also used, along with the gills, in respiratory exchange of oxygen – thus the system acts in a similar way to a blood circulation. All this is co-ordinated by a complex nervous system, including a sensory system which 'tells' the sea urchin what the conditions are. This is poorly understood at present. At least some of the echinoids have light-sensitive areas near the tube-feet pores.

As mentioned, the first echinoids appear in the fossil record in the Ordovician period, around 490 million years ago. These Perischoechinoide persist more or less alone amongst the echinoids until the Triassic period (245– 208 million years ago). At this point there begin to appear other varieties of echinoids – all of them variations of the basic structure that appeared back in the Ordovician. Many became extinct but others continued to the present and include the cidaroids (indistinguishable from the original Ordovician stock), sea urchins, heart urchins and sand dollars. The differences between these echinoids are to do with the shape and use of the existing organs which were already fully present in the original stock 490 million years ago. This seems a good case of microevolution – small changes which accumulate over time and which can produce different species, all with the same overall structure. There is nothing much 'new' when we look at the millions of years of change that has occurred.

If we try to look in detail at the changes that occur over limited stretches of time we can see micro-evolution under the microscope, so to speak. We are talking here of the changes seen over tens rather than hundreds of millions of years. A good example is the study done by Rowe[8] of a class of echinoid known as Micraster. He looked at the changes seen in the fossils of this echinoid during the latter part of the Cretaceous period (which ended 65 million years ago). What we see are minor, albeit significant, changes including the following: broadening of the test, deepening of the anterior groove (this groove is characteristic of this class), movement of the mouth anteriorly and lengthening of the ambulacra (ambulacra are rows of calcitic plates which are part of the skeleton). This is the sort of study that shows what actually happens over time in an organism. There are no new organs or structures. The variations that micro-evolution produces can be explained on the basis of variations in distribution of existing genes - the subject of a later chapter.

There is a continuing debate amongst palaeontologists as to the rate at which evolution occurs. The gradualist school sees evolution as a gradual and progressive phenomenon – similar to what Darwin believed. The 'punctuated equilibrium' school sees change as occurring in sudden bursts. Punctuated equilibrium is the term coined by Eldredge and Gould in 1973[9] when they described the fossil record, as they and many see it, as showing stasis most of the time with sudden dramatic changes every so often. They postulated that the species as a whole can be 'selected' in natural selection, not just individuals – this accounts for what they see as sudden, punctuated changes. Some believe that the change leading to new species usually occurs in peripheral, isolated populations which do not easily make it into the geological record – even if the changes occur over thousands of generations. The new species then migrate into the area where they proliferate and where they are seen in the record as a sudden new type.

In fact, in contrast to the work of Rowe, described above, the general situation found in successive strata is that of stasis. A recent paper by McCormick and Fortey[10] studied the trilobite *Carolinites* as it progressed through the Lower to Middle Ordovician period (many millions of years). What they saw was essentially a confirmation of the idea that stasis is the norm. Changes were seen over time in various characteristics of the trilobite. Some changes in anatomy were sudden, some seemed to go through successive transitional forms and some fluctuated with little or no change. Those changes that occurred showed evidence of reversal – that is, the changes sometimes reverted back to the original. Statistical analysis of the entire process showed that there was no evidence of any sustained 'direction' in any changes.

Looking at this gradualism vs. punctuated equilibrium debate as someone outside the palaeontological profession can be daunting. What can we make of it? It seems clear that what is being debated mainly is not macro-evolution but micro-evolution. The entire question about punctuated equilibrium seems to be mainly about the 'micro' scale. I have already stated that micro-evolution is a fact (small changes in existing organisms). What is not in any way explained, however, are the big changes that we call macro-evolution. These are the sudden appearance of completely new forms and structures – indeed, all the appearances of the phyla and classes (higher taxa). These remain a total mystery and do not fit in with any known mechanism or theory. Now more technical stuff.

Staying with the echinoderms, what about the other forms? We will look briefly at the principle groups. There are (at least) four subphyla of the main echinoderm phylum. These include: the Echinozoa, which contains the classes of sea urchins just described and also the sea cucumber class; the Asterozoa, which contains the classes of starfish and brittle stars; the Crinozoa, which contains the class of sea lilies; and the Blastozoa, an extinct group. Sea cucumbers appear in the Ordovician period. Starfish and brittle stars appear in the early Ordovician and sea lilies appear around the same time (though there are Crinozoa, similar to sea lilies, in the early Cambrian). There are however many forms of extinct echinoderms found in the early Cambrian period. The earliest is known as Helicoplacoidea and it is clearly an echinoderm, fusiform in shape with plates similar to modern ones. The echinoderm phylum illustrates, in fact, the wide early disparity that I mentioned in Chapter 2. To quote Stephen Gould again[11]:

'Palaeontologists have long recognised the Burgess pattern of max-imal early disparity in conventional groups of fossils with hard parts. The echinoderms provide our premier example. All modern representatives of this exclusively marine phylum fall into five major groups – the starfishes (Asteroidea), the brittle stars (Ophiuroidea), the sea urchins and sand dollars (Echinoidea), the sea-lilies (Crinoidea), and the sea-cucumbers (Holothuridea). All share the basic pattern of fivefold radial symmetry. Yet Lower Palaeozoic rocks, at the inception of the phylum, house some twen-ty to thirty basic groups of echinoderms, including some anatomies far outside the modern boundaries. The Edrioasteroids built their globular skeletons in three-part symmetry. The bilateral symmetry of some 'carpoids' is so pronounced that a few palaeontologists view them as possible ancestors of fishes, and therefore of us as well. The bizarre helicoplacoids grew just a single food groove (not five), wound about the skeleton in a screwlike spiral. None of these groups survived the Palaeozoic, and all modern echinoderms occupy the restricted realm of five-part symmetry. Yet none of these ancient groups show any sign of anatomical insufficiency, or any hint of elimination by competition from surviving designs.'[11]

Clearly the unravelling of the relationships between the various echinoderms is highly complicated and all that I have written is just a summary. This is what makes the study of fossils very much a specialist area which is totally inaccessible to the ordinary enquirer. That is a pity because the fossil record is something which should be available to all - it is the evidence of the origin of life and must be in the public domain. My hope is that what I write here can begin to clarify the main facts amongst the huge amount of information.

Let me quote again the authoritative palaeontologist E.N.K.Clarkson: 'The earliest echinoderms were already highly diversified on their first appearance in the Lower Cambrian.'[12]

Later, he writes: 'Even in the Lower Cambrian these early echinoderms are widely separated in morphology and their relationships can only be understood cladistically.'

Cladistics, as already mentioned, is one way of classifiying organisms according to shared characteristics; as such it can conveniently dispense with actual evidence based on fossils lineages. Suffice it to say that there are no transitional forms found in the fossil record that link the various types of echinoderm since their first appearance in the Cambrian period. What we see is the initial phylum appearing suddenly around 530 million years ago – in a wide variety of forms. Subsequently, we see the fivefold radial symmetry surviving in the major groups that appear later. I have searched hard for any evidence of transitional forms between the major groups which exist today and I have seen none. What we do see is small changes in existing groups (micro-evolution). The initial phylum appears along with all the others in the Cambrian explosion, it is fully complex and has more classes than we have now, the classes that survive appear suddenly also. The only clear evolution is on the micro scale. It is the sudden appearance of new forms without linking transitionals that remains the real mystery – just as it was in Darwin's day.

All that is written about the echinoderms is mirrored in the fossil record of the other invertebrates. I have used the echinoderms as one example of what is found also amongst fossils of the other main phyla of invertebrates: arthropods, molluscs, sponges, cnidarians, bryozoans and brachiopods. Check any textbook on the subject and you will see that sudden appearance and discontinuity is the norm. The problem that Darwin so clearly saw has not gone away with time and the fossil record seems now to refute the entire basis of his theory of gradual changes. In the following chapter we will look in detail at the fossil record of the vertebrates.

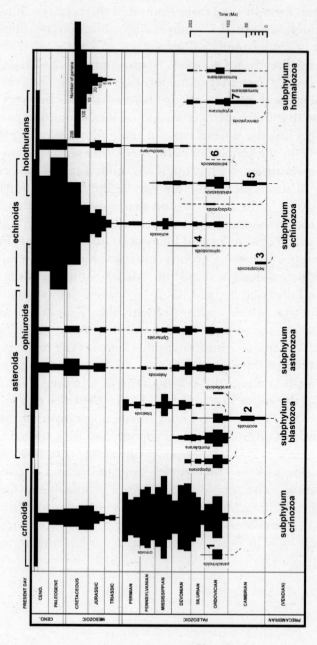

figure 4.2

Time ranges and abundance of echinoderm groups. (based on Sprinkle 1983).

Chapter Four

1. Charles Darwin (1859). *The Origin of Species.* John Muarry
2. M.A. Norell and M.J. Novacek (1992). 'The fossil record and evolution: comparing cladistic and palaeontological evidence for vertebrate history', *Science,* 255PP, 1690–3.
3. M.J. Benton and R. Hitchin (1996). 'Testing the quality of the fossil record by groups and by major habitats', *Historical Biology,* 12,PP 111–57.
4. Michael Benton, (1997). *Vertebrate Palaeontology.* Chapman and Hall.
5. ENK Clarkson (1993). *Invertebrate Palaeontology and Evolution.* 3rd edin. Chapman and Hall. P.45
6. Helpful is: Janet Moore (2001). *An Introduction to the Invertebrates.* Cambridge University Press.
7. Clarkson, *Invertebrate Palaeontology and Evolution,* p302.
8. A.W. Rowe, (1899). 'An analysis of the genus Micraster, as determined by rigid zonal collecting from the zone of Rhynchonella cuvieri to that of Micraster coranguinum', *Quarterly Journal of the Geological Society of London* 55PP, 494–547.
9 N.Eldredge and S.J. Gould, (1973). 'Punctuated Equilibria: An Alternative to Phyletic Gradualism', in *Models in Paleobiology* ed. TJM Schopf. San Francisco: Freeman, Cooper and Co., pp. 82–115.
10. T. McCormick and R. Fortey,. (2002). 'The Ordovician Trilobite *Carolinites,* a test case for microevolution in a macrofossil lineage', *Palaeontology,* vol. 45. part 2.
11 Stephen Jay Gould, (1990). *Wonderful Life. The Burgess Shale and the Nature of History.* New York: Vantage, PP 301 – 302.
12. Clarkson, *Invertebrate Palaeontology and Evolution,* P.302

Chapter Five

The Fossil Record of Vertebrates

The origin of vertebrates

Vertebrates consist of all animals with backbones; that is fish, amphibians, reptiles, birds and mammals. Vertebrates are themselves classified as members of a larger group or phylum known as the chordates.

The chordates all have a notochord (a firm rod of cells) running longitudinally in the body, which is represented by the vertebrae in vertebrates. They also have typical muscle bands. The oldest fossil chordates known are fish from the Lower Cambrian of China, in the Chengjiang deposit.[1] We have already seen *Pikaea* in the Burgess Shale – an early Cambrian eel-like chordate. Of the living chordates, the closest in relationship to us vertebrates are said to be *Amphioxus*, a small eel-like sand burrower, and (somewhat amazingly) the larval stages of sea squirts.

Until the recent finds in Chengjiang it was thought that the oldest known fossil vertebrate was *Anatolepis* – a small fish of the late Cambrian (520–505 million years ago). It is identified as a vertebrate because of the presence of apatite (calcium phosphate) bone and also dentine.

The origin of the chordates is as mysterious as the origin of any known phylum. There are no intermediate transitional fossils to link them to any other invertebrate. Benton writes:

'It is hard to find any reason for pairing adults of the phylum Chordata with any particular group of worms, molluscs, arthropods, or other potential relatives.'[2]

When an animal embryo develops from a single fertilised cell, the cell divides repeatedly until a hollow ball of cells is formed known as the blastula. A pocket of cells then moves inwards from the outside (as if one indents a tennis ball) and the opening of this deep pocket is called the blastopore. In most invertebrates (the *protostomes*) the blastopore becomes the mouth, while in others (the *deuterostomes*), including the chordates, this opening becomes the anus – and the mouth is a secondary perforation. Evolutionary theory claims that the *deuterostomes* evolved from the *protostomes*. All chordates are deuterostomes and it is necessary for evolutionists to envisage an extraordinarily unlikely process – that of completely changing this basic embryological structure from forming the mouth to forming the anus. Benton writes:

'Such a dramatic turn-around, a switch from mouth to anus, seems incredible, and its evolution is a mystery of course since there are no intermediate stages!'

The exclamation mark and the word *incredible* in the above quote are Benton's - this is just one more example of the puzzlement amongst palaeontologists about the lack of transitional forms and the seeming impossibility of one form evolving into another. There is also much dispute about the evolutionary relationships among the deuterostomes and there are no intermediate fossils to help us unravel how the vertebrates arose from any other group. There is even a considerable body of opinion that we may have arisen from the sea squirt tadpole by a process called paedomorphosis: the idea is that a sea squirt tadpole 540 million years ago failed to mature to adulthood and initiated a new class of animal. There is no direct evidence for this and the whole debate about the various relationships seems to be somewhat chaotic and lacking in coherence.

In this chapter I will outline just a small selection of developments seen over time in the fossil record of vertebrates to give a reasonable picture of the evidence we have. The reader who wants to know more should consult a textbook on palaeontology. The appearance of humans will be covered in a separate chapter.

The fossil record of fishes

The first vertebrates were fish. The very earliest fish, as mentioned above, are found in the Lower Cambrian. They only appear more abundantly in the Ordovician (510–439 million years ago) and all the major groups were present by the Silurian/early Devonian.

The earliest fish were jawless (of the class *Agnatha*). By the Ordovician there were several different distinct groups of these, many of them heavily armoured with bony head shields and plates over the body. The only surviving jawless fish now are the lampreys and hagfish.

The Ordovician and Silurian periods saw the emergence of various groups of fish with jaws (*Gnathostomes*). The classical theory is that jaws formed from modified anterior gill arches of jawless fish. The idea is that the anterior three to four arches (made of cartilage or bone) somehow fused to form the complex jaw bones (with teeth) attached to the skull. As so often in evolutionary theory, however, the reality is rarely so simple or plausible. As Benton (honest as always) says in his textbook:

> 'Some anatomical evidence, however, suggests that the gill-arch theory may not be so simple in reality. The agnathan gill arches are externally placed, and not homologous with

the internal gill arches of gnathostomes, so that jaws arose after the transition from external to internal gill arches. In addition, the palatoquadrate and Meckel's cartilage [the upper and lower jaws] may never have been gill arches at all, but were always associated with the mouth from their first appearance.'[2]

Basically, what he is saying is that the jaws could not have arisen from the gill arches because the jawed fish have entirely different gill arches (internal) and these internal arches must have been there before the appearance of jaws. There are, as is becoming predictable, no transitional fossils to help us see any intermediate structures. Once again, in a vitally important anatomical change, there is no evidence of the gradual progression needed by Darwinists.

During the Silurian and Devonian periods there appear a wide variety of jawed fish, including the *Placoderm* class – fish with bony carapaces over head and shoulders. The class became extinct by the end of the Devonian. The first sharks appear in the Devonian. Sharks and rays are cartilaginous (without bony skeletons) and are thought to be the most primitive living fish. *Cladoselache*, one of the first sharks, looks remarkably like one of our modern versions.

True bony fish appear as various groups around the early Devonian period. They are divided into those with ray fins (*Actinopterygii*) and lobe fins (*Sarcopterygii*). The ray-finned fish are the dominant type in the sea today. The lobe-finned type includes the living lungfish and the coelacanth. Lungfish have lungs as well as gills and use these when pools become stagnant. They are able to 'haul' themselves to fresh pools by use of their fins.

The coelacanth is a lobe-finned fish that had been seen fossilised from the middle Devonian to the late Cretaceous. It was thought to have become extinct then and was a prime candidate for being an ancestor of land animals. In 1938, a fish was caught off the coast of Africa which turned out to be a coelacanth (though now called *Latimeria*). Numerous others have been caught since – and as such it is a 'living fossil'. We see here complete stasis for 100 million years. The *Latimeria* is essentially the same as the coelocanth. Not only this, but when dissected, the *Latimeria* shows no evidence at all of the numerous anatomical adaptations needed to survive on land. When a living coelacanth is observed it is seen to use its lobe fins purely for water paddling – there is no hint of 'walking' on the sea bottom. This, of course, is not proof against some other lobe finned fish gradually developing the apparatus of terrestrial life but it does show us that organisms can remain unchanged indefinitely. Evolving is not apparently in the coelacanth's interests.

We will discuss the anatomy of fins in more detail later when tackling the subject of development of legs for use on land. Current evolutionary theory sees the lobe fin as a precursor of the limbs of amphibians.

A typical textbook of palaeontology will show a diagram (see example, figure 5.1) of the appearance over time of the various fish groups.

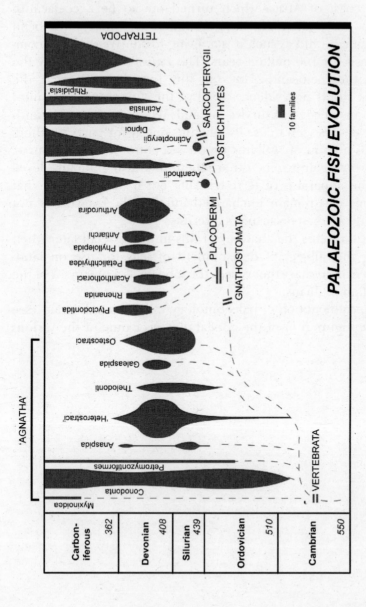

figure 5.1

The evolution of early and mid-Palaeozoic fishes. The pattern of relationships (indicated with dashed lines) is based on cladograms described above, and the 'balloon' shapes indicate the span in time of known fossils (vertical axis) and their relative diversity (horizontal scale)

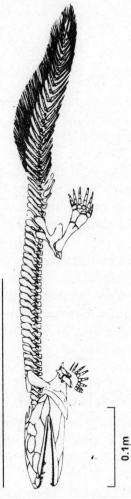

a) *Acanthostega*

0.1m

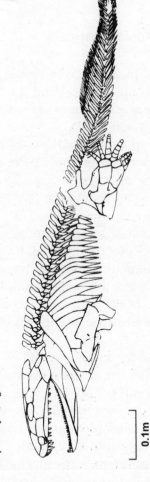

b) *Ichthyostega*

0.1m

figure 5.2

Reconstructions of The Late Devonian tetrapods Acanthostega and Ichthyostega.

As before, we see distinct types of organism with hypothetical dotted lines linking them – supposedly showing the actual descent paths. However, there are no transitional fossils to complete those dotted lines.

The first tetrapods

The first terrestrial vertebrates, able to walk on land, are found in the Devonian period and are known as tetrapods. The earliest specimens are thought to be *Acanthostega* and *Ichthyostega* (see diagrams in figure 5.2).

To explain the transition from water to land the evolutionist must try to imagine what the gains were. Many thought that it could be as a result of pools drying up during droughts, thus encouraging changes in fish that would allow them to move on the shore to other pools. The problem with this is that the adaptations needed were for returning to water, not for living on the land. It is now thought that the enormous adaptations seen in the first tetrapods could not have been for just moving from dried pool to water. Instead it is now believed that the move to land was because of the available food on the shore line; that is the plants and the already existent terrestrial invertebrates.

Pause for a moment and try to imagine the situation. Fish are in the water near the shore and there is food available on land if the fish can get out and walk (as well as breathe). As I detail the changes needed for such a transition, try to imagine how the multitude of presumed intermediate stages (on the way to becoming tetrapods) could have fared.

We must remember that for Darwinian evolution to work there must be a clear survival advantage for the organism at every stage in its transition from one form to another. Every small change, presumed to be caused by random mutations, will only continue to be reproduced if the change causes an improvement. Remember too that the fish is already beautifully adapted to living in the water, where there is food available. There had to be some great advantage for it to abandon such an aquatic existence. Why should it do so? How could the first mutant fish that had half-fin and half-leg actually fare better than its cousins in the water? As we shall see, for any sort of reasonable life on land there had to be some extraordinary changes in fish anatomy and physiology but (and I emphasise this) it is hard to see how any intermediate stages from fin to legs could have had any survival advantage. Many palaeontologists feel that Acanthostega was only able to paddle in shallow water or on top vegetation – rather than actually take its full weight on land. This does seem conjectural though and the Acanthostega limb is very different from the fin of a fish. There are no fossil intermediate stages of half fin/half-leg but evolutionary theory has to believe that there were. Let us look at the details.

A fish is supported by water and its bodyweight is effectively zero. On land, however, the body is held up by limbs. As a result of this the entire internal anatomy and skeleton of the tetrapod has to be changed to cope with the forces of gravity. For example, the vertebrae and their muscles had to become adapted to prevent the body sagging between the limbs.

The ancestral, lobe-finned fish is thought to be of a group known as the *osteolepiforms* and a prime candidate now is one called *Eusthenopteron*. The pectoral fin of the fish is thought to have evolved into the forelimb of the first tetrapod (see diagrams in figure 5.3).

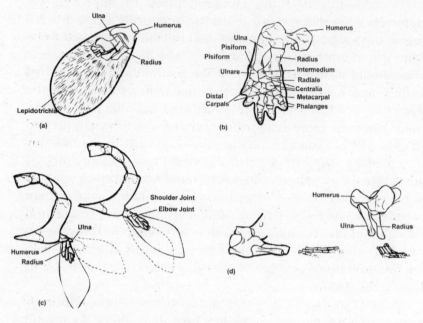

figure 5.3
The origin of tetrapod limbs and walking on land: (a) pectoral fin of the
osteolepiform fish *Eusthenopteron* showing postulated identities of the
bones; (b) equivalent forelimb of the basal tetrapod *Eryops*;
(c) possible movements of the forelimb of *Eusthenopteron*; (d) step cycle
of the forelimb of the primitive amphibian *Proterogyrinus*. (after various
sources).

The collection of bones at the base of the fin has been
described as comparable with the basic bones found in the tetrapod limb – that is humerus, ulna and radius. Such a connection
is, however, fairly speculative – given that the shapes of the bones
in the tetrapod limb are completely different, some of the lobe-
fin bones do not get into the tetrapod limb and, most importantly, the tetrapod limb has a complicated arrangement of new wrist
bones (ulnare, radiale and intermedium) as well as carpals,

metacarpals and phalanges. The same problems exist for imagining the transition of the fish pelvic (rear) fin to hind limb.

In order to become a tetrapod limb, the lobe fin required new bones, new joints and completely new musculature and limb girdle. The limbs had to be used in an entirely different way from fins – requiring altered orientation and completely new neuro-muscular co-ordination. The pectoral girdle in the fish is attached to the skull and that of the first and subsequent tetrapods is well behind the skull. The pelvic girdle in the fish is a small unit embedded in the body wall whereas that of the tetra-pod is firmly attached to the vertebral column. The jaw hinge of the first tetrapod is also quite different from that of a fish.

The transition to air breathing required the development of efficient lungs. It is postulated that the fish ancestor of tetrapods had lungs already – similar to the modern lungfish. There had to be loss of extra, unneeded fins (*Eusthenopteron* had two dorsal fins, an anal fin and a large tail fin). There also had to be a major change in the method of reproduction – it is postulated that the earliest tetrapods laid eggs in water and had an initial tadpole-like stage like our modern amphibians. Remember that this is quite unlike any stage in fish development and that to acquire an initial aquatic tadpole form which then developed limbs and emerged on land, is a huge macro leap in biological function.

Try again to think through these changes, which all appear suddenly in the fossil record, and imagine how they could all have been harmonised, by entirely chance mutations, to give us the tetrapod on land that we see abruptly appearing in the fossil record. I believe that, once again, the Darwinist model fails us here.

Continuity of form

I have been greatly helped in my search for the fish to tetrapod evidence by Jennifer Clack's excellent book *Gaining Ground. The Origin and Evolution of Tetrapods.*[3] By looking in detail at the known fossils of the earliest tetrapods we can definitely see continuity between the lobe-finned fish precursors and those first land walkers. Any creationist must take this into account. When we look at either *Acanthostega* or *Ichthyostega* we see the following features which definitely show that they are based on a fish anatomy: they retain a sort of fish shape, they apparently retained internal gills, they retain a lateral sensory line on the body (as in fish) and they have a sort of tail fin – albeit of a different shape. Here is continuity amidst the discontinuity of the other new features such as the limbs. We see here that, even though there is an unexplained leap to terrestrial locomotion, there is linkage with fish. Later, when discussing homologies in Chapter 12, we shall see that such continuity is common in different groups of animals. This does not validate Darwinism, however. These are not 'transitional' in the sense that Darwin meant and longed for. We see too much that is suddenly new to call the first tetrapods transitional. There is no gradual evolution here but we do see earlier forms being a sort of template for the creation of later forms. The appearance of tetrapods is sudden and is just one of the saltations (leaps of form) that characterise the fossil record.

An interesting point, though, is that these earliest land walkers do not have the pentadactyl (five-digit) limb pattern of all future tetrapods. *Acanthostega* has been found to have had eight fingers and *Ichthyostega* had seven toes. Later tetrapods seem to have had variable numbers of digits, from six to eight. It is not until the early Carboniferous period that we find the pentadactyl limb in tetrapods.

We need to find at least some common ground between creationists and palaeontologists on these matters. Creationists generally will not tolerate anything that smacks of a transitional fossil – the features described for the first tetrapods might, to the uninitiated, seem transitional and so they are just denied by some creationists.

I well remember one television documentary which showed a creationist (Duane Gish) and a palaeontologist (Per Ahlberg, a world authority on fish/tetrapod fossils). They were interviewed separately – which is just as well. Ahlberg was clearly incensed by any idea of creation of each form instantly, without intermediates. He was using his expertise to try to confound Gish, who was firmly in the group that refuses to see any transitionals. More heat than light ensued.

This book will hopefully give a truer picture than either camp. I hope that I will have demonstrated that there are indeed totally unexplained yawning gaps in the fossil record but that there is also continuity between succeeding forms – hence the fish-like characteristics of the first tetrapods. They appear with all the attributes of land walkers, suddenly – but retain signs of their lineage. Darwinism requires smooth continuity always. We do not see this. There have been enough fossil beds examined (particularly in Greenland) for a clear picture of the fish to tetrapod evidence. The transition to tetrapod is sudden. Darwinists will put this down to inadequate fossil preservation but this argument is now wearing thin. The challenge to the Darwinist is to show us even one case in the history of life where a macro change has occurred smoothly with well- defined, gradually transitional intermediate forms. Darwin waited and we still wait.

The appearance of birds.

In 1861 some men who were working in a quarry near Pappenheim, in Bavaria, came across a well-preserved fossil of a creature with outspread wings. This was the now-famous Upper Sonhofen Lithographic limestone – used for making printing blocks. The men brought the fossil to the local medical officer, a Dr Friederich Karl Haeberlein, who was a collector of plant and animal fossils. He then sold it to the British Museum where it was examined by the renowned anatomist Richard Owen. The fossil appeared just after the publication of Darwin's *Origin of Species* and it seemed to provide concrete evidence of a transitional form between reptiles and birds. As such it was used by Thomas Huxley and others to champion Darwin's theory.

Since then there have been six other fossils of *Archaeopteryx* found in the same limestone. All are from the late Jurassic period of about 150 million years ago and, as such, are the earliest discovered bird. *Archaeopteryx* was about the size of a magpie and had fully developed wings with feathers. These feathers were identical to those on modern birds. Some of the specimens are truely exquisite, the best one being the *Berlin* specimen of 1877 (see figure 5.4).

It is thought that the environment was a sort of tropical lagoon and many other marine organisms are found in the sediments along with various terrestrial animals and plants. The *Archaeopteryx* birds are thought to have floated on the water after death, sunk to the bottom after the intestines had burst and become submerged in sediment quite intact. They are fully articulated with head bent back because of the contraction of the neck muscles after death.

The reason that many think this to be a true transitional form is because Archaeopteryx has features that are also found in reptiles and dinosaurs but not in modern birds. In particular it has a long bony tail, teeth in both jaws and in its wing there are three separate clawed 'fingers'. It is also seen that the sternum (breast bone) is not keeled as in modern birds (which have a large keel like projection for attachment of flight muscles). There is no doubt that it is a remarkable and unusual bird and it is understandable that palaeontologists have seen it as evidence for evolution. This is the first reaction most of us could have when we see *Archaeopteryx*.

Once more we need to pause before swallowing whole an initial impression which may be based more on our preconceptions

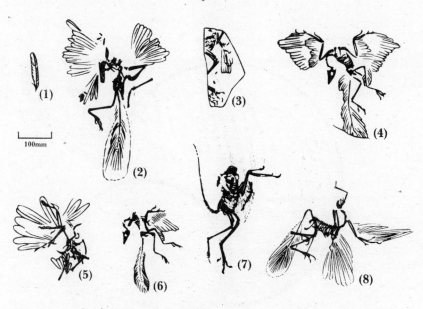

figure 5.4
The specimens of Archaeopteryx, all drawn to the same scale, with the bones shown in black, and the feathers in rough outline. The commonly used specimen names, and dates of discovery are as follows: (1) Berlin/Munchen 1860; (2) London 1861; (3) Haarlem 1855 (1970); (4) Berlin 1877; (5) Maxberg 1956; (6) Eichstatt 1951; (7) Solnhofen 1987; (8) Solnhofer Aktien–Verein 1992. (after Wellnhofer, 1988b, 1993.)

than real logic. The main problem to be solved is how did such a bird (even if unusual and seemingly transitional) get to evolve flight and feathers? Few doubt that this animal could fly because the wings are fully developed and it has typical feathers of flying birds (as opposed to flightless ones), because the vanes of the feathers are asymmetrical and the feathers are curved – both adaptations for flight. It also has fused clavicles, or a 'wishbone', only found in birds and its skull and bones are specially light – again only seen in birds as an adaptation for flight.

Feathers are remarkable structures (see figure 5.5).

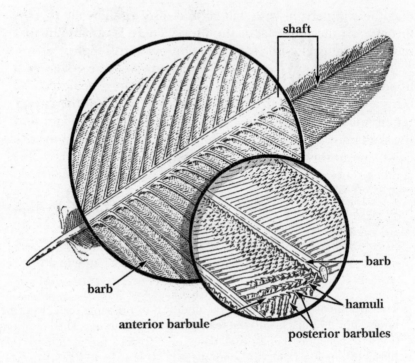

figure 5.5
(from Tyne and Berger)

Each feather has a shaft carrying an array of barbs positioned at right angles to the shaft to form the vane. These barbs are held together by a series of barbules and from the anterior barbules hooks project downwards to interlock with ridges on the posterior barbules. A feather has around 1 million barbules and they work together to form an impervious vane. Flight feathers are amazingly light and yet strong. The feathers are made to provide very sophisticated aerodynamics for flight. By letting part of the air stream escape between slots in the feather, turbulence is reduced. Tendons, in many birds, intricately twist the feathers so that the wing shape can change and also to allow air to pass through on the upstroke of the wings. There is no aircraft that remotely approaches the sophistication of a bird's flight.

There are many changes in the anatomy and physiology of a hopeful dinosaur that must occur before flight is achieved. All birds are warm blooded, a necessary precondition for the energy requirements of flight. Bones are hollow to reduce weight and the skull must be very light and thin. The heart must be more efficient, and to this end birds have a four-chambered heart, unlike reptiles. The lungs need to be enlarged and more efficient and birds have a unique system of air flow in the lungs which allows better gas exchange – the air flows through the lung as opposed to just in and out. The brain is modified considerably to co-ordinate flight. Perhaps the most amazing thing is the actual development of wings along with the muscles and their attachments.

All this may seem like just another creationist diatribe extolling the wonders of nature without understanding how evolution could have produced it all over long periods of time. But we really must be serious here in our attempt to imagine what steps could conceivably have occurred to bring about *Archaeopteryx* from some land-bound tetrapod by gradual change. Remember again that every single evolutionary step must confer some advantage to the organism in survival or fitness for that organism to continue. Candidates for the ancestor to the birds include mammals, crododilomorphs, a group called archosaurs and lastly dinosaurs. The latter is currently the favourite.

Once again the fossil record does not help us here. There are, so far, no fossils that show any intermediate stage between a dinosaur and *Archaeopteryx*. (I discuss fossils of feathered dinosaurs, from a period later than *Archaeopteryx* below.) This is becoming the holy grail of fossil hunters – so much so that any fossil that does turn up to link birds and dinosaurs is worth a lot of money. Of course, in these situations human greed gets to work and recently a cleverly constructed fake was sold for a lot of money and sent to America from China. It looked like a dinosaur with bird features including feathers and it was (astonishingly) accepted by many experts as a real transitional fossil. The *National Geographic* magazine made a front page story about this fossil, as it often does when any piece of evolutionary evidence is turned up. Soon it was discovered that the fossil was, in reality, a composite of different fossils glued together. I mention this not to pour scorn on scientists who, being human, do make errors – but to show how desperate scientists are to find this 'missing link' which continues to evade them.

Since I began to write this book there have been further finds of feathered fossils from China. Right now, there is an exhibition of some of these in the Natural History Museum in London. The counter slab (the opposite side of the fossil slab) of the hoax just mentioned is said to have feathers genuinely as part of the original dinosaur. Another recent find, *Sinosauropteryx* has a skeleton similar to a meat-eating dinosaur and has feather-like down appearing to cover its body (presumably for insulation). It does not have flight feathers, however. The latest sensation is an extraordinary animal which has feathers, including, it is claimed, flight feathers on four limbs (except they cannot be for flight!).

The main problem for the Darwinist with these finds is that they come from a fossil bed that is considerably younger than the oldest *Archaeopteryx* fossil. The oldest *Archaeopteryx* is 147 million years old, whereas these Chinese ones are all around 120 million years old. They therefore cannot be the ancestors of Archaeopteryx.

Palaeontologists are therefore in some dispute. Larry Martin of the University of Kansas is very sceptical and feels that these fossils may be modified descendants of *Archaeopteryx*. Others (perhaps the majority) feel that these Chinese fossils shared a common ancestor with *Archaeopteryx* – in the same way that modern apes are supposed to share an ancestor with us. What is very clear is that they are not ancestral to the first birds and therefore we have no right to assume anything. There are times when a respectful restraint should be shown. Inevitably, the latest *National Geographic* has a large feature on the new find, proclaiming it as a likely ancestor of modern birds.

Essentially there are two schools of thought about how dinosaurs became birds: ground up or arboreal (gliding down from trees). Neither idea has won the day yet but we have to imagine each scenario as best we can to see if it is possible by current evolutionary thinking. I have read various ideas in textbooks but have not found any that do not seem very contrived indeed.

The ground up school must try to envisage a dinosaur developing rather shaggy scales and 'flapping' along, perhaps to catch insects or to try going faster. It seems quite inconceivable to me that scales have developed by accident (which we must remember is what mutations are) into the magnificent feathers we see on *Archaeopteryx*. Clearly the initial changes to the scales could not have been for flight because flight was not possible then – so how did the modifications for flight get chosen? If the changed scales were initially for insulation, for example, then how did they become used in flight which is an entirely different function? We really should expect here definite intermediaries in the fossil record to show us how this could happen – let alone all the other adaptations needed for flight which I have mentioned.

The 'gliding from trees' or arboreal school must see how a dinosaur developed the ability, in stages, to glide/jump from on high. Again, we have to understand the necessary sequence. Initially actual flight would be impossible, so the creature would need to half-jump and half-glide downwards – presumably as a form of escape or to catch food. The problem with gliding is that the physical adaptations needed for it are in direct opposition to those of powered flight. Also, once again, the 'perfection' of the feather for powered flight leads us to doubt any such gradual and circuitous process. I recommend a good discussion about all of these problems found in Michael Denton's book *Evolution: a Theory in Crisis*.[4]

I have just read the most recent book by the experts in this area (a collection of papers, entitled *Feathered Dragons[5]*). It does not really add anything to these arguments. There seems much speculation (the feather impressions in the fossils are very hard to see in many cases and the rather outrageous illustrations betray considerable artistic licence).

Let us go back to Darwin who wrote in his *Origin of Species:*
If it could be demonstrated that any complex organ existed which could not possibly have been formed by numerous, successive, slight modifications, my theory would absolutely break down.

Darwinists must take these words very seriously because we are now at the stage when an accumulation of facts does in fact lead us to severely doubt that any such gradual process has ever produced complex organs. This is backed up by our knowledge about the inability of mutations to bring about such macro changes - to be discussed later.

The story of the horse

The evolution of the horse has become over the years a classic, almost iconic, sequence which textbooks illustrate as a gradual progression over millions of years. It is perhaps the most paraded evidence for evolution that we see (figure 5.6).

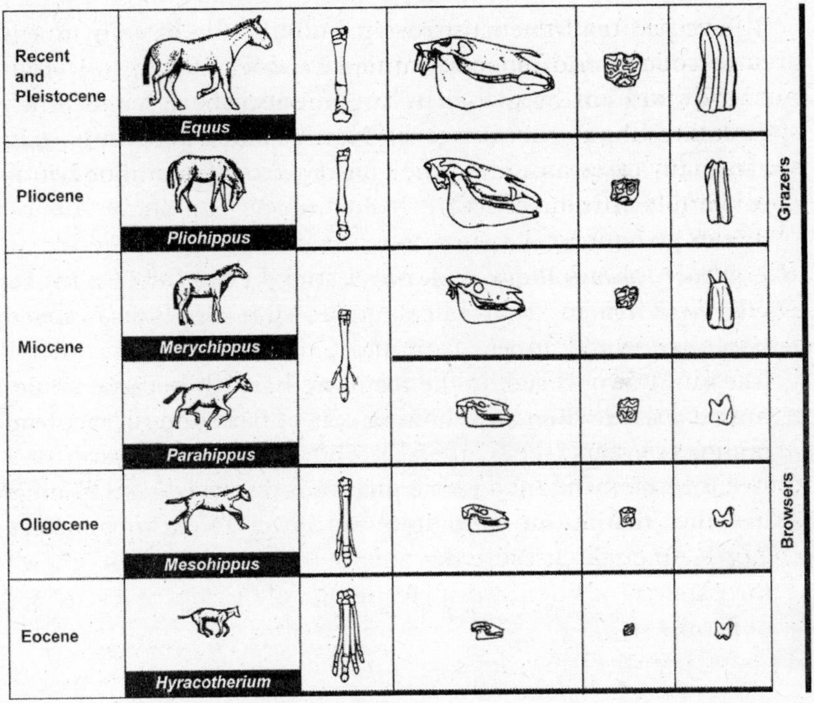

Recent and Pleistocene	Equus					Grazers
Pliocene	Pliohippus					
Miocene	Merychippus					
	Parahippus					
Oligocene	Mesohippus					Browsers
Eocene	Hyracotherium					

figure 5.6

What is the truth of the matter? The main candidate for a horse ancestor is *Hydracotherium*, a terrier-sized animal of the early Eocene (about 50 million years ago). It is unclear what relationship this animal had with any precursors. *Hydracotherium* had four toes on the front and three on the back feet. A series of fossils is found which seems to show gradual changes leading to the modern horse *Equus*. By the Miocene period *Mesohippus* had just three toes in front and later we see *Pliohippus* in the Pliocene period with only one toe. This leads on to *Equus*, which looks similar to *Pliohippus* but larger. Alongside the reduction in toe number we see increase in body size and a change in the teeth from leaf-crushing molars to more deep-rooted, tougher teeth with complex infoldings suited to grass grinding. It is thought that the development and radiation of more modern forms was because of the spread of grasslands in North America.

Interestingly, the horses of North America became extinct about 10,000 years ago. In the seventeenth century modern horses were reintroduced there from Europe (to where American horses had migrated).

Nearly always there is shown a straight line of descent from *Hydracotherium* to *Equus*. In fact, the fossil record is far more complex than would appear from most accounts. If a true ancestral tree was drawn, based on the fossils we have, then there would be numerous side branches showing other coexistent forms over the millions of years (see figure 5.7). The actual line of descent is not so easily seen when the whole picture is displayed. It is found, for instance, that at one time there were over 16 existing species in North America. Is there not at least some wishful thinking when the experts confidently display the line of descent as a clear series of forms?

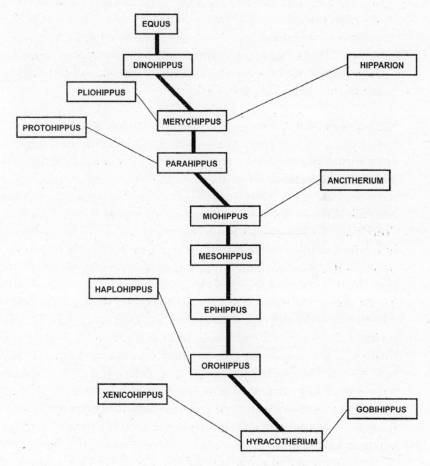

figure 5.7
A more recent outline of horse evolution

To quote Michael Benton again:

'The story of the horse has become a text-book example of 'progressive evolution' or a 'trend' since there seems to be a clear-cut one-way line of change from the small, leaf-eating Hydracotherium to the large, grazing Equus. However, there is no evidence for uniform change, and the pattern of evolution is rather more complex than it might at first seem. There was no single line of evolution from Hydracotherium to Equus, and many side-lines branched off in the Oligocene and Miocene.'[6]

Having said that, there is evidence that ancestors of horses did have extra toes – in particular we see this when, rarely, a horse is born with extra toes – a presumed genetic remnant of the past and there is evidence of vestigial digits on modern horses. What is more important, however, is to examine what we are describing carefully. Is this a case of macro-or micro-evolution? There is every reason to assume it is micro-evolution. As will be discussed in a later chapter, micro-evolution is a process that is not in dispute but it cannot explain any major new structures or forms. We do not see in the horse series any major change in anatomy, just loss of digits, increased size and altered tooth shape. This hardly constitutes evidence for Darwinian evolution, except on the micro scale. We need only to look at the variety of shapes and forms of, for example, dogs to see how selective breeding (in this case directed by man) can lead to very different sizes and shapes – even in one species. We know, however, that the variety of dogs is not because of new genetic information; it is because of shuffling the existing genes (I discuss dog varieties in more detail in Chapter 15).

It is worth mentioning just a little about the appearance of plant groups around the same time as the development of vertebrates. The flowering plants, for example, appear suddenly and dramatically in the fossil record with no pre-cursor forms that lead in to them. Jane Francis, Professor of Palaeo-climatology at the University of Leeds, said the following on the BBC Radio 4 programme ' In our time' on 17th February 2005:

"The flowering plants originated about 100 million years ago, so it is a much younger period of rocks (than the Burgess shale which was being discussed) and we have good fossils there so we can study things in a lot more detail. And we have been able to work out these and we can see the early ancestors of them and we have a very good record. But the interesting thing is we don't see what we think is the very early flowering plants. You know, there's a big gap and we have suspicions that there may be 20, 30, 40 million years when the flowering plants evolved and we just have no evidence of them".[7]

Once again we see sudden leaps in form and complexity which are not explicable in a Darwinian framework.

The evidence we have of the history of vertebrates runs counter to the ideas of Darwin. We may see small micro changes over time but the big ones come suddenly, without transitional forms. We are left again without any clear mechanism for the variety of forms that abound amongst this group of animals.

1. Simon Conway Morris. Personal communication.

2. Michael Benton (1997). *Vertebrate Palaeontology*. Chapman and Hall.

3. Jennifer Clack (2002). *Gaining Ground. The Origin and Evolution of Tetrapods*. Indiana: Indiana University. Press.

4. Michael Denton (1986). *Evolution – a Theory in Crisis* Adler & Adler.

5 Philip J Currie, Eva B.Koppelhus, Martini A.Shugar, Joanna L.Wright (2004). *Feathered Dragons. Studies on the Transition from Dinosaurs to Birds*. Indiana: Indiana University Press.

6. Benton, *Vertebrate Palaeontology*, PP. 342-343

7. Jane Francis. BBC Radio 4, 'In our time' February 2005.

Chapter Six

Human Origins

Any proper treatment of the theory of evolution must include a thorough investigation of the evidence for the origin of *Homo sapiens*.

I will first attempt to reconstruct briefly the current prevailing beliefs about our origins which are held by most anthropologists.[1]

Molecular evidence

Molecular phylogeneticists use techniques to compare the proteins and DNA of humans and modern apes. By comparing these they try to calculate how far back their common ancestor was. The idea is based upon the belief that mutations in our DNA occur at a steady rate. Therefore, the more differences there are between the DNA of two species, the further back in time was the split from a common ancestor.

When we look at the DNA of humans, African apes (gorillas and chimps) and orang-utans we see that human DNA differs least from chimpanzees, then gorillas and is most distant from the orang-utan. This has led to the belief that both humans and chimps have evolved from a common ancestor, with the ancestor of gorillas splitting off the lineage earlier and the orang-utan ancestors having split off earlier still. It is widely proclaimed that our DNA is 98 per cent the same as that of chimpanzees. This begs the question as to how the remaining 2 per cent can account for such amazing differences. The 98 per cent refers to known genes. In my chapter on genetics I explain that human DNA is mostly sequences that have no known use – sometimes referred to as 'junk' DNA. I discuss there the very recent findings that such 'junk' has been found to have profound uses within the cell. This research, therefore, throws doubt on whether the 98 per cent figure is true. The fact that humans share 60% of our genes with bananas does not make us 60% banana – so the close genetic similarity we have to chimps needs to be put in perspective.

Fossil evidence is said to date our common ancestor with orang-utans to be 30 million years ago. By knowing the genetic distance between us and orang-utans (comparing DNA) it is therefore possible to calibrate the so-called molecular clock. The molecular clock is the idea mentioned above, that the mutations causing divergence of bases in the DNA occur regularly and average out at a steady rate.

It was then found that the genetic distance between humans and orang-utans is six times the distance between that of humans and chimpanzees. From this a simple calculation gives the figure of 5 million years since the divergence of hominids from African apes. In fact, this figure has since been pushed back to about 7.5 million years ago.

Most people are aware that these figures are approximate (plus or minus 1 or 2 million years) and not everybody accepts the validity of the molecular clock. We cannot be sure that mutations in DNA have always occurred at the same rate. There are confounding factors, such as the fact that some parts of DNA mutate faster than others. Also, the longer the time interval is, the more likely it is to get more than one mutation at the same site – perhaps even reversing the first mutation.

Furthermore, the evidence that we have of a continuous lineage stretching back 30 million years to a common ancestor with orang-utans is flimsy, there being so few fossils and so many gaps, as I will show later. So the very foundation of the calculation to calibrate this molecular clock depends entirely on a prior belief in our common ancestry with orang-utans. If that is removed then we have no calculation. Having said this, the work done is impressive and holds sway very strongly in most scientific circles.

The fossil evidence

So, the hominids are thought to have evolved from an African ape ancestor around 7.5 million years ago. It is known that around this time there were climatic and geological changes which led to deforestation in East Africa. It is postulated that bipedalism, the upright stance of hominids, evolved to allow them to exploit the clear areas. Bipedalism, though less efficient than quadruped locomotion, is more efficient than the crouching, knuckle-using gait of the ape.

Every account of the fossil evidence for early hominids, if honest, continually declares the frustrating poverty of material. It has been said that if all the fragments of early hominids were put together they would just about fill a suitcase. The earliest specimens (which are controversial) include Orrorin consisting of fragments of 5 individuals found in Kenya dated about 6 million years old. We then have to wait about 2.25 million years to get any further substantial evidence; this is the famous set of footprints found at Laetoli in Tanzania in 1974 by Mary Leakey. Study of the prints seems to indicate bipedalism.

A portion of skull was found in 2002 in the Djurab desert in Chad by Michel Brunet of the University of Poitiers.[2] He claims this is the earliest hominid as it is almost 7 million years old. Called *Sahelanthropos tchadensis*, the only evidence that it could be our ancestor seems to be that it has somewhat shorter canines than are found on apes. Otherwise, it appears to be an ape. Many, if not most, experts do not agree with Brunet about the significance of the find.[3] Flimsy evidence such as this is very commonly put forward in the intense race to succeed.

There have been various finds of 'hominids' in Africa from around 4.4 million years ago. The earliest of these seems to be *Ardipithecus ramidus*. The fragments of this creature include part of a child's mandible, a part of the base of a skull and the bones of an individual's arm. The position of the attachment of the vertebral column to the skull is considered to be indicative of partial bipedalism. There are then some fragments of a creature named *Australopithecus anamensis* dated at about 4.2 million years ago. The next available specimen, chronologically, is the better known as *Australopithecus afarensis*. The most famous example is that of *'Lucy'*. Lucy, dated 3 million years old, was discovered in Ethiopia in 1974 and is a partially complete skeleton.

This specimen, the later 1.6 million-year-old *'Turkana boy'* (*Homo erectus species*) and the partial skeleton of 'Nariokotome boy' (*Homo ergaster* dated 1.6 million years also), are the only partially complete skeletons of hominids ever found up to the appearance of Neanderthal man (150,000 years ago). It is important to keep this in mind when weighing up the rather confident claims of anthropologists about the evidence before us. All we have otherwise are fragments and the occasional more complete cranium.

There is not much to distinguish Lucy from apes apart from her supposed bipedalism. The bipedalism itself has been shown to be quite suspect. Careful studies by Professor Randall Susman of State University of New York[4] have shown that reconstructed models of Lucy would not have been able to walk differently from apes. Not only this but the toe bones in Lucy are curved, as in ape toes – indicating a tree-climbing anatomy. Therefore, the evidence for bipedalism in *Australopithecus afarensis* just does not exist despite very confident claims. There are, however, changes to more human-like teeth – to the extent of displaying relatively small canines, a thick enamel layer and cheek teeth better designed for grinding tough material than for dealing with fruit. All these are thought to be indications of divergence from ape-like to more human-like anatomy. There are many primitive – that is ape-like – features about her, including a small brain, protruding incisors and ape-like premolars. The tooth row seems inbetween the U-shape of an ape and the parabolic arch of a modern human, being more V-shaped. The overall appearance of the head is ape-like, with a small brain and a prognathic (protruding) jaw. It is hard, therefore, to agree that Lucy was a hominid.

At this point it is worth inserting the account of *Ramapithecus*. During the 1960s and 1970s the majority of anthropologists believed that the hominids had arisen at least 15 million years ago. This was because of the fragmentary fossil evidence of a species known as *Ramapithecus* found at sites in Europe, Asia and Africa. The oldest specimen was 15 million years old. Fossil remains of *Ramapithecus* were first found in 1932 in northern India by G. Edward Lewis, a graduate of Yale University. The specimen consisted of two fragments of an upper jaw. Lewis wrote a paper that claimed this was an early hominid. This was not accepted by the establishment until 1961, when Elwyn Simons, of Yale, re-examined the fossils and declared that Lewis had been correct. He based this upon the shape of the jaw and the configuration of the teeth. He reconstructed the jaw (which was missing a central section) and declared it to be arched, as in humans. There were other indicators of more human-like features, including a short non-projecting jaw, small canines, human-like molars, thick enamel and differential wear on the molars (a feature found in humans because of the different times of eruption of these teeth).

The first biochemical evidence of hominid origins occurring only 5 million years ago was published in 1967. This made anthropologists re-examine the jaw of *Ramapithecus* and they came to the conclusion that the jaw was more V-shaped than the human arch and that thick enamel was apparently found in some apes (including orang-utan). *Ramapithecus* was consigned, then, to being just another Miocene ape, closely related to another known as

Sivapithecus. As *Sivapithecus* is considered to be closely related to the orang-utan, and hominids were, by then, considered to be more related to African apes, *Ramapithecus* was out ruled as our ancestor.

I relate the story of *Ramapithecus* for two reasons. Firstly, it shows the continual flux of ideas about our origins which continues to this day. Secondly, it will be noted that the jaw of *Ramapithecus* did, in fact, have features suggestive of being hominid – in particular the V-shaped arch (found in *Australopithecus afarensis*), the small canines, the short face and the human-like molars. All these features are used now to claim our relationship with *Australopithecus afarensis* and yet in *Ramapithecus* they have come to be considered as purely ape-like. Such inconsistency does not encourage us to be confident about our relationship to the early 'hominid' finds.

The fossil record of fragments and craniums following the appearance of *Australopithecus afarensis* indicates that three to four other species of Australopithecines followed. From one of these, *Australopithecus africanus*, it is thought that the first member of the genus *Homo* sprang, in Africa, around 2.5 million years ago. This coincides with the first appearance of stone tools.

Homo habilis, the first member of the genus, was distinguished from the Australopithecines by having a larger brain of between 600 and 800 ccs; as opposed to the Australopithecine 450 ccs. A modern human has an average brain size of 1,350 ccs. *Homo habilis* also had smaller cheek teeth and larger anterior teeth, was bipedal and was much more slender in build than the Australopithecines, as well as being taller.

For about 1.5 million years the Australopithecines existed alongside the *Homo* genus. The last Australopithecine is thought to have been just over 1 million years ago. Before this extinction, there was another development; the appearance of another species of hominid, *Homo erectus*, about 1.5 million years ago.

Homo erectus, which is thought to have evolved from *Homo habilis*, had a slightly larger brain of 900 ccs and had changes in facial features, including the development of prominent ridges above the eyes and quite a flat face. Some anthropologists call the earlier forms of *Homo erectus* found in Africa by the name *Homo ergaster*. *Homo erectus* bones are very much the same shape and size as those of modern humans but are much more thickly buttressed.

The overall anatomy of *Homo erectus* did not change in its 1.2 million-year history apart from the head. By the time it was to become extinct about 300,000 years ago, the brain size was 1,100 ccs. It is not known whether this was a gradual or step-like increase as the number of fossils is so few. The later, larger-brained members of the species are sometimes known as *Archaic sapiens*. The appearance of *Homo erectus* coincides with an advance in the form of stone tools, known as the Acheulian technology. The principal difference between this and the older Oldowan technology was the addition of several larger implements, including the hand axe, the cleaver and the pick. The hallmark of the technology is the hand axe, a teardrop-shaped tool that was much more difficult to make than the Oldowan assemblages. As far as other culture or technology is concerned, we have no indication from archaeology, and there are no carvings, paintings or jewellery associated with *Homo erectus*.

Homo erectus is thought to have spread from Africa into the 'old world' around 1 million years ago, because after this time we see fossils of this species in Europe, the Middle East and Asia. *Archaic sapiens*, the supposed end-point of *Homo erectus* evolution, is said to have lived from about 0.5–1.0 million years ago. There is debate as to whether the evolution of the species occurred in one place or whether it arose simultaneously in different parts of the world. There is also controversy as to whether the specimens of *Homo erectus* in Africa and Asia are the same or different species. Interestingly, the tool technology of the Asian *Homo erectus* was of the old Oldowan type. As I will explain below, there is also controversy as to whether there was a smooth, direct line evolution from *Homo erectus* to *Homo sapiens* or whether it included some other intermediaries.

In 2004 there was a sensation when a partial skeleton, including skull, of a very small hominid was found on the island of Flores in south-east Asia. This has been claimed to represent a descendant of *Homo erectus* and lived just 18,000 years ago. Modern *Homo sapiens* lived alongside this creature. It appears that there is dispute however amongst the experts, some of whom claim that the skull is more likely from a human, possibly suffering from a condition called microcephaly.

Neanderthal man

In 1856, three years before the publication of Darwin's *Origin of Species*, some quarry workers in the Neander valley, near the Dussel river in Germany, unearthed some bones which appeared human. The site was not properly excavated but they recovered the top of a cranium, some leg and arm bones, and other damaged pieces. They were taken to a local teacher and historian, Carl Fuhlrott. He thought that they had some unusual features – particularly the thickness of the bones and the heavy build compared with modern humans. He enlisted the help of Hermann Schaafhausen, Professor of Anatomy at the University of Bonn. He agreed that the bones belonged to some primitive form of human and together they formally described the find to the scientific community. There followed a controversy that still continues, to some extent this day; was Neanderthal man a direct ancestor of modern man or was he an evolutionary side-shoot which became extinct?

For some years, Neanderthal man was confined to obscurity because in the 1870s, Rudolf Virchow, the German anatomist and pathologist, pronounced the bones to be modern and diseased, not ancient. By the end of the nineteenth century, however, there were more and more finds of the species (there are about 200 individuals found to date). The scientific community began to accept that these were not human bones.

The Neanderthals have distinct differences from modern humans, including a larger brain (around 1,500 ccs). The skull is long and low and, particularly from the side, has a very different shape. They lived in Europe and the Middle East between 150,000 and 34,000 years ago. There is some evidence that they buried their dead and in one case may have put flowers around the dead person. Pollen of such flowers was found in the soil of an apparent grave belonging to a Neanderthal, buried 60,000 years ago in Shanidar, Northern Iraq.

Neanderthals lived in the same region and time as modern humans who appear in various parts of Europe and the Middle East between 100,000 and 40,000 years ago (before Neanderthals disappeared). It is very hard to know whether or how they interacted or even possibly interbred.

Where do modern humans, *Homo sapiens*, come into the picture? Therein lies much controversy – something that we have already found out is extremely common in this business. There have been many theories over the past few years and they include: first, a direct line evolving from the Neanderthals; the second, known as the spectrum hypothesis, varied this by suggesting greater interaction between different populations, which blurred the distinction between them; third, the pre-Neanderthal model whereby an early Neanderthal population diverged relatively recently to produce the 'classic' Neanderthals of western Europe – which became extinct – and modern humans; fourth is the *pre-sapiens* model whereby an ancient split is envisaged between the true *sapiens* lineage leading to modern humans and the lineage leading to other groups such as *Homo erectus* and the Neanderthals; fifth is the idea that modern humans had a single centre of origin and then migrated into the rest of the 'old world'; Sixth, the hypothesis that *Homo erectus* evolved in multiple regions, with genetic mixing between them, into modern humans – known as the multiregional model. The last two models are now the most popular with the dominant one being the fifth – a single origin of modern humans in Africa around 200,000 years ago, who gradually migrated to populate the rest of the 'old world', reaching Europe between 100,000 and 40,000 years ago. There is evidence of modern human remains in Australia from 60,000 years ago – presumed to have reached there by sea.

The idea that we are directly descended from Neanderthals has now been more or less thrown out by the amazing extraction and analysis of DNA from the original Neanderthal specimen found in 1856. In 1997 this was found to have significant differences in the sequence of nucleotide bases from DNA in humans.

Tools and artefacts

A particular stone tool technique, known as the Levallois technique, appeared 250,000 years ago. It allowed the manufacture of many flakes of a standard shape and size, superseding the Acheulian technology. This technique coincides with the appearance of *Archaic sapiens* the larger-brained version of Homo erectus. Strangely, there is no change in this technology until 30,000–40,000 years ago when the Upper stone-age culture is found. This culture had many more types of tool and the very first ornaments, jewellery and cave paintings (although some interesting stones with regular markings have been found in a cave in Southern Africa, dated around 70,000 years ago, – it is disputed whether these are significant evidence for advanced humanity).

It remains a mystery as to why there are no earlier remnants of this modern culture, considering the fact that anatomically 'modern' *Homo sapiens* existed much earlier. Could it be that somewhere along the line we are confusing gross anatomy and brain size with something that cannot be measured in bones – *humanity*?

Brain size in itself is a poor indicator of intelligence amongst mammals. An elephant's brain is five times bigger than humans and weighs about 8 kilograms. One might, however, think that it is the percentage of body weight that is important. The elephant brain is 0.2 per cent of its bodyweight compared to the human brain, which is 2.33 per cent of bodyweight. However, this is not the whole story either – a shrew's brain is 3.33% of its bodyweight yet no one suggests it is more intelligent that humans. We therefore need to take with a pinch of salt the suggestion that the increasing brain sizes amongst primate fossils is exactly equivalent to an increase in intelligence.[5]

The idea that modern humans emerged from Africa has received a boost from the comparison of DNA from mitochondria. Mitochondrial DNA is outside the cell nucleus and as such is only passed on through the maternal line. This is because the mother's ovum contains all the extra nuclear material of the fertilised cell. By complicated analysis and comparisons of mitochondrial DNA from large groups of different races, it is found that there are more variations between the nucleotide sequences of African people's mitochondrial DNA than for other races. Based on the same 'molecular clock' idea, this suggests that the earliest modern humans were African. A possible date is also put on this origin – approximately 200,000 years ago.

Those who support the multiregional origin of humans hotly dispute all of this, especially Milford Wolpoff of the University of Michigan. Others dispute the place of origin and suggest the Middle East as also a possibility (an intriguing thought for the upholder of Genesis). The data from the mitochondrial DNA studies indicate the original human stock to have been a small group. (A group of two is not ruled out.)

Piltdown man

The saga of Piltdown man needs to be mentioned briefly – if only as a salutary warning[6] Charles Dawson (1864–1916) was an amateur archaeologist who recovered some pieces of human skull from a gravel pit near Piltdown Common in the south of England between 1908 and 1912. Sir Arthur Smith Woodward of the British Museum thought the bones to be important and joined Dawson in further searches. A fragment of a jaw was then found from the same site and this was thought to belong to the same individual. The jaw was ape-like but the skull was like that of a human. This combination of an ancient jaw and a large brain fitted a prevailing theory for the very early origin of man, with Neanderthals being a primitive offshoot. The British anthropological establishment accepted the find and included Sir Arthur Keith, one of the most prominent British anthropologists of the early twentieth century. Many, particularly in America, did not agree and were suspicious.

It was not until 1949, 40 years later, that the specimen was found to be a forgery. It was a modern skull with the jaw of an orang-utan, which had been stained with dye.

This story is not related in order to gloat at sincere experts who 'got it wrong'. It does warn us though that even the top experts can be totally misled by something which, in hindsight of course, was a fairly crude forgery. Key to this, I believe, is the fact that the forgery fitted the prevailing theories and so it was accepted. The question is: how often are anthropologists now also trying to 'fit' fossil fragments into their particular theory?

It will hopefully be evident from this brief outline that the further we go back, the more uncertainty there is. Look at the raging controversies and constant changes of opinion there are about our most recent origins, despite having far more fossils of Neanderthals than for all the early hominids put together – then you have to agree that any categorical conclusions about direct ancestry are quite unscientific. There are about 200 Neanderthal individuals found – and it has taken DNA testing of a specimen to put to rest the idea that we are directly descended from them. We have no DNA from *Australopithecus afarensis* and only a few fragments of bone and yet many anthropologists are assuming they are our direct ancestors.

Deep time

Part of the problem is that of 'deep time'. This is the title of a recent book by the palaeontologist and editor of the journal *Nature*, Henry Gee.[7] He reminds us of our inability as humans to get our heads around the immensity of millions of years. He shows us how ridiculous it is to unearth a piece of cranium dated 3 million years old and then confidently place it in any lineage at all, let alone ours. He writes:

> 'The intervals of time that separate fossils are so huge that we cannot say anything definite about their possible connection through ancestry and descent.'

He also says that each fossil is:

> 'an isolated point, with no knowable connection to any other given fossil, and all float around in an overwhelming sea of gaps.'

He points out that all evidence for human evolution

> 'between about 10 and 5 million years ago – several thousand generations of living creatures – can be fitted into a small box.' He concludes: 'To take a line of fossils and claim that they represent a lineage is not a scientific hypothesis

that can be tested, but an assertion that carries the same validity as a bedtime story – amusing, perhaps even instructive, but not scientific.'[8]

Gee is no creationist and is very much of the scientific establishment. He has been courageous enough, however, to express the doubts that many honest palaeontologists have.

The emergence of Homo sapiens

An important meeting of the British Academy and the Academy of Medical Sciences was held in 2002. The meeting was to discuss and present papers about the 'speciation of modern *Homo sapiens*'. It brought together some of the world's top experts in palaeontology, archaeology, linguistics, neurology and genetics. The proceedings of this meeting are helpfully available in book form[9]. What comes across immediately is the mystery of how *Homo sapiens* arose so suddenly. It is generally recognised that the real distinguishing feature of humans from all other life is the ability to communicate in very complex language. It is felt widely that this is linked with other forms of expression such as art and the way in which we are uniquely self-conscious. The differences in brain and especially language ability between humans and any other mammal are huge. How did we acquire such complex brains and particularly the ability to communicate in language - so suddenly?

Darwin, in his book *The Descent of Man*, felt that, like all other forms of evolution, the ability to use language must have come gradually.[10] He, like modern Darwinists, could not cope with any idea of saltation or sudden evolutionary jumps. Saltations (large macro jumps in evolution), as I have explained elsewhere, imply some other mechanism than natural selection. He could not see humanity as any different ultimately from other animals as far as evolution is concerned. Here he differed from his fellow evolutionist A.R.Wallace, who felt that man came about by some other agency. Wallace said (in contrast to Darwin):

'My view on the other hand, was, and is, that there is a difference in kind, intellectually and morally, between man and other animals; and that while his body was undoubtedly developed by the continuous modification of some ancestral animal form, some different agency, analogous to that which first produced organic life, and then originated consciousness, came into play in order to develop the higher intellectual and spiritual nature of man.'[11]

Wallace here is strangely up to date, because amongst many leaders in the field of anthropology and palaeontology, there is great puzzlement as to how man arose so quickly with so many differences from apes – particularly as regards language.

Tim J. Crow writes about the problem of how our species became isolated from all the other hominids surrounding it over the millennia.[12] A crucial concept of speciation is that there is reproductive (usually also geographic) isolation so that a new species can form. Hominids seem, however, to have lived side by side. How did the species come about? What environmental factors directed the isolation of our forebears from the other hominids? On this there is no obvious answer.

But, getting back to language, there is perplexity amongst experts as to how it could have just arrived by gradual processes. F. M. Muller, who held the chair of Philology at Oxford University, delivered three lectures to the Royal Institution in 1873[13]. In these he severely criticised Darwin's bland assertion that human language just evolved gradually – as if there was some continuity between animals and humans. The problem is that the level and sophistication of language that we possess, along with all the hardware it needs, is immense if put alongside any other creature. This includes an extremely complex neo-cortex with special language/speech area, a multitude of connections within the brain and to the voice apparatus, a new and refined type of larynx and vocal tract, to say nothing of the auditory processes that enable us to interpret and think about the sounds. In his second lecture Muller says:

'There is one difficulty which Mr Darwin has not sufficiently appreciated. ... There is between the whole animal kingdom on the one side, and man, even in his lowest state, on the other, a barrier which no animal has ever crossed; and that barrier is – Language ... If anything has a right to the name of specific difference from our philosophic dictionaries, I should still hold that nothing deserves the name of man except what is able to speak.'

Muller then acknowledges that Darwin had conceded this point of difference with animals. He quotes Darwin from *The Descent of Man*. It is not the mere power of articulation that distinguishes man from other animals, for as everyone knows, parrots can talk; but it is his large power of connecting *definite sounds with definite ideas*.[11] Muller writes:

'Here, then, we might again imagine that Mr Darwin admitted all that we want, viz. that some kind of language is peculiar to man, and distinguishes man from the other animals ... but, no, there follows immediately ...'This obviously depends upon the development of the mental faculties.'

Muller asks:

'What can be the meaning of this sentence? ... If it refers to the mental faculties of man, then no doubt it may be said to be obvious. But if it is meant to refer to the mental faculties of the gorilla, then whether it be true or not, it is, at all events, so far from being obvious, that the very opposite might be called so - I mean the fact that no development of the mental faculties has ever enabled one single animal to connect one single definite idea with one single definite word.

I confess that after reading again and again what Mr Darwin has written on the subject of language; I cannot understand how he could bring himself to sum up the subject as follows: 'We have seen that the faculty of articulate speech in itself does not offer any insuperable objection to the belief that man has been developed from some lower animal'.[14]

Muller distinguishes between emotional and rational language. The former is more onomatopoeic – 'the power of showing by outer signs what we feel, or it may be, what we think'.[18] He feels that this is shared by man with animals. However, rational language he relates to the power of forming and handling general concepts, and this is regarded as specific to man. Diseases that affect the anterior lobe of the left side of the brain interfere with this advanced type of language. It is the ability to form 'roots' that he regards as the essence of rational language. He describes this as follows:

'There is in every language a certain layer of words which may be called purely emotional. It is smaller or larger according to the genius of each nation, but it is never quite concealed by the alter strata of rational speech. Most interjections, mainly imitative words, belong to this class. They are perfectly clear in their origin, and it could never be maintained that they rest on general concepts. But if we deduct that inorganic substratum, all the rest of language, whether among ourselves or among the lowest barbarians, can be traced back to roots, and every one of these roots is a sign of a general concept. ... Take any word you like, trace it back historically to its most primitive form, and you will find that besides the derivative elements, which can easily be separated, it contains a predicative root, and that in this predicative root rests all the connotive power of the word.'

Muller summarised this:

'If the words of our language could be derived from imitative or interjectional sounds, such as bow-wow or pooh-pooh, then I should say that Hume was right against Kant, and that Mr Darwin was right in representing the change of animal into human language as a mere question of time. If on the other contrary ... after deducting the purely onomatopoeic portion of the dictionary, the real bulk of the language is derived from roots, definite in their form and general in their meaning, then that period in the history of language which gave rise to these roots ... forms the frontier ... between man and beast ... That period may have been of slow growth, or it may have been an instantaneous evolution: we do not know ... These roots, which are in reality our oldest title-deeds as rational beings, still supply the living sap of the millions of words scattered over the globe, [*while no trace of them, or anything corresponding to them, has ever been discovered even amongst the most advanced of catarrhine apes.*] (My emphasis.)

Linguists, neurologists and others have tried to imagine the way in which evolution could have produced, so quickly, the hard-wiring needed to produce the complexity of language (combined with the changes in vocalisition and auditory processing needed). Stephen Pinker writes:

'Language ... is composed of many parts: syntax, with its discrete combinatorial system building words; a capricious lexicon; a revamped vocal tract; phonological rules and structures; speech perception; parsing algorithms; learning algorithms. Those parts are physically realized as intricately structured neural circuits, laid down by a cascade of precisely timed genetic events.'[15]

T. J. Crow questions this statement and asks: if these genetic events are sequential innovations, at what stage and in what order were they introduced? He argues that there seems no clear way in which such huge changes could have occurred gradually, somehow in the transition between *Homo erectus* and modern *Homo sapiens.* It actually comes down to the problem of irreducible complexity that has been discussed elsewhere at various points in this book. The wiring and structure needed for our language skills is so perfectly harmonised and complete in itself that even the hardened Darwinist baulks at the amount of chance mutations needed in short time to bring it all together. T. J. Crow (the editor of the Proceedings of the recent British Academy meeting of world experts on this subject) and others have therefore seriously entertained the old-fashioned concept of some saltationary event – a huge macro mutation that somehow conjured it up in one great unlikely leap. By doing so they immediately remove themselves from any real Darwinian position and into the realm of mystery. There is no understanding of how the sort of huge macro event needed could have happened by pure chance. The sceptical reader should read the afore-mentioned book edited by T.J.Crow, to have confirmed what I am saying here about a saltation being held responsible for our appearance.

Pinker, in his afore-mentioned book *The Language Instinct,* confirms the findings of Chomsky that humans of all races are born with the hard-wiring of their brains to deal with the most complex grammar and that this is a 'universal' grammar which is found in every language. I cannot resist quoting Pinker here because, although a Darwinist, he cannot avoid the use of the term 'design' in his admiration of the complexity involved:

> 'At the very least I hope you are impressed at how syntax is a Darwinian "organ of extreme perfection and complication." Syntax is complex, but the complexity is there for a reason. For our thoughts are surely even more complex, and we are limited by a mouth that can pronounce a single word at a time. Science has begun to crack the beautifully designed code that our brains use to convey complex thoughts as words and their orderings.'

I hope that this somewhat extended section on the acquisition of language has shown that, amongst those who are most knowledgeable, there is considerable perplexity as to how any Darwinian model could arrive at *Homo sapiens*.

We do not know where we originated as a species – perhaps Africa, perhaps the Middle East. We do not know when and how we emerged but what evidence we have indicates a massive saltational leap from the preceding *Homo erectus* and astounding changes in a brain which triples in size within just 3 million years. There is absolutely no evidence for gradual change. It is true that the fossil record shows a sort of progression that appears to be ape – *Australopithecus* – *Homo habilis* – *Homo erectus* – *Archaic sapiens* – human. It is understandable that the experts want to see this as a neat ancestry but we are not able to go back in time to see the changes actually happening, 'deep time' forbids us to make lineages out of a suitcase of bones, and we still have no idea why and how modern humans appeared.

Having looked at the available evidence as coldly as I can manage, I firmly believe that the arrival of *Homo sapiens* was indeed a saltational event. As I have explained previously, a saltation is when a completely new form comes on the scene without any clear explanation. It is not a gradual change as is needed in the Darwinian model but a full-scale novelty of a complexity which defies any known genetic mechanism. Chance mutations are not the answer because the probability of, for example, arriving at the language ability of humans from an ape precursor is far far too unlikely. This does not out rule some prior connection with the apes but it does out rule any known natural process in bringing about that change.

This saltational event, which characterised our appearance, fits in with the fossil record for other organisms. All the major novelties appear abruptly.

Are we descended from apes or not? It is clear that we are not in the sense that Darwin meant. It has definitely not been a question of gradual evolution. Saltations require something more than chance and selection; they require design and purpose.

1. Much material is from Roger Lewin (1998). *The Origin of Modern Humans*. *Scientific American* Library.

2. Michael Brunet et al (2002). 'A new hominid from the upper Miocene of Chad, Central Africa', *Nature* pp. 418, 145–151.

3 Kate Wong (2003). 'An ancestor to call our own. New look at human evolution', *Scientific American*, special edn.

4. Randall Susman, along with Jack Stern in Open University module on evolution.

5. See Susan Greenfield (1997). *The Human Brain*. Phoenix.

6. See: Walsh, John E. (1996). *Unravelling Piltdown*. Random House.

7 Henry Gee (2000). *Deep Time*. Fourth Estate.

8. Ibid.

9. T.J. Crow (ed.) (2002). *The Speciation of Modern Homo sapiens. Proceedings of the British Academy*. Oxford: Oxford University. Press.

10. Charles Darwin (1871). *The Descent of Man and Selection in Relation to Sex*. Princeton: Princeton University Press.

11. A.R. Wallace, 1905, quoted in Crow, *The Speciation of Modern Homo sapiens,*

12. Crow, *The Speciaton of Modern Homo sapiens.*

13. F.M Muller (1996). 'Lectures on Mr Darwin's philosophy of language', in *The Origins of Lanaguage* (ed. R. Harris), Bristol: Thoemmes Press pp 147–233

14. *Ibid.,*

15. S. Pinker (1994). *The Language Instinct*

Chapter Seven

The Genetics of Darwinism

When Darwin lived and developed his theory he had no idea at all what the mechanism of variation and inheritance was. He knew that organisms inherit the characteristics of parents but his idea was that the process involved a form of blending of fluids from each parent to produce a mixture of their qualities in the progeny. The problem with this idea is that it assumes a dilution of each quality by mixing from both parents and the logical outcome might be a uniform and homogeneous similarity after many generations. Darwin was also somewhat Lamarckian in his ideas. Lamarck had put forward the theory that an animal or any organism could influence the nature of its offspring by the way in which it lived. For example, a giraffe might, by repeatedly stretching its neck, somehow pass on to its offspring a tendency for longer necks. We now know that changes to the somatic cells of the body, which are not involved in reproduction, have no effect on the offspring. However, it is understandable that, without knowing the basis for inheritance, which we now grasp, it was easy for scientists to propose such views.

Little did Darwin know that in the 1860s an Augustinian monk by the name of Gregor Mendel was performing a set of experiments that pointed to the existence of biological elements that we now call genes. Mendel was born in Moravia, part of the Austro-Hungarian Empire. After high school he entered the Augustinian monastery of St Thomas in the city of Brunn, now called Brno, in the Czech Republic. The monastery was dedicated to teaching science and doing scientific research and so Mendel was sent to a university in Vienna to obtain his teaching credentials. He failed his examinations and returned to the monastery. He then started on his experiments with plant breeding, which have posthumously earned him the title of founder of genetics.

Mendel's work is a superb example of good science. He chose the garden pea (*Pisum satvium*) to perform a series of beautifully thought out breeding experiments, looking at the inheritance of seven easily measured variables found in the pea. These were characters such as round or wrinkled peas, purple or white petals, and long and short stems. By crossing pea plants which were pure bred for these characters, he was able to work out the pattern of inheritance of each character in the offspring of future generations. By assiduous and patient work he managed to analyse mathematically how characters appear in the generations. By his work he was able to show that the inheritance of characters was not that of *blending* (as Darwin thought) but was *particulate*. The various characters, when expressed in any plant, were not mixtures but retained their exact nature. He came to understand that the part of the parent which passed on any character was a gene (though he did not coin this term). He was also able to show that each plant had genes, which came in pairs, and that each parent contributed one of its paired genes to the offspring. He found that when the genes from each parent (say for colour of petals) were expressed in the offspring, then some genes were dominant and others recessive. If a dominant gene was combined with a recessive gene then the character of the dominant gene was expressed in the plant.

Mendel's work was put into a single paper entitled 'Experiments on Plant Hybridisation' and presented in 1865 to the Brunn Natural History Society. It was 35 years before any notice was taken of it by the scientific community.

In 1902 Walter Sutton, an American, and Theodor Boveri, a German, independently recognised that Mendel's inherited particles during the production of gametes (male and female sex cells) parallels the behaviour of chromosomes at meiosis (the division of cells in the formation of gametes). Genes are in pairs, as are chromosomes; the alleles of a gene (the different forms of gene that occur at a locus on the chromosome) segregate equally into gametes, as do the members of paired chromosomes, and different genes act independently, as do different chromosome pairs. Sutton and Boveri concluded from this that genes are located on chromosomes.

Speciation and micro-evolution

All this work led in stages to our understanding of genetic inheritance, leading in 1953 to the discovery by Crick and Watson of the molecular structure of DNA, the molecule that chromosomes are made of. We now know that genes are, in fact, sections of this thread-like double helical molecule. We also understand the answer to what makes a species. Cats always have kittens just as people always have babies. It is the genes that dictate the properties of each species. The product of most genes is protein and proteins are the main macro-molecules of an organism. What you see when you observe a cat is either protein or something that has been made by protein, and the amino acid sequence of a protein is encoded in a gene.

We can also understand how we get variation within a species. No two cats (or humans) are quite the same. This is because each gene can exist in several forms that differ from one another – usually in small ways. These forms of genes are the alleles that have been mentioned. Each individual (apart from identical twins) has a unique set of alleles that makes the individual distinctive.

It was now possible to really work out a plausible mechanism for evolutionary change to occur. Darwinism basically is natural selection acting on variation. In any given environment some members of the species survive and reproduce better than others and so these are selected to pass on their genes. For example, a cheetah that is faster may well have better biological *fitness* than its slower siblings. Thus, it is more likely to reach adulthood and pass on its genes.

Within any population of a species there is variety which derives from different combinations of alleles. At any locus on a chromosome, a gene, as I have said, may come in more than one variety and it is this variety which is said to fuel the evolution of new forms. In the case of Mendel's peas, for example, the gene that codes for colour of flower (purple or white) will have a ratio of combinations in the whole population. In this case we can call the allele for purple C and that for white c. C is written in upper case because it is dominant. The possible combinations of alleles are: C/C, C/c and c/c. The first two are expressed as purple and the last is white. The colours, which are the physical expression of the genes in the organism, are part of the plant's *phenotype* (as opposed to the *genotype*, which is the combination of genes of that organism).

If you have time to read a textbook on evolution (such as the one I have just read from the Open University)[1], you will find that much space is devoted to trying to work out the way in which new species can form from an original parent population. Nearly all of this study is to do with the combinations of alleles that occur in a population. It is widely understood that new species usually form because a population has become split so that two or more parts of it are geographically separated (for example, by a mountain range or by a continent dividing). By being separated physically there can be a difference in the frequency of alleles between the two populations – whereas if they remain together then constant interbreeding will tend to lead to a fixed ratio or combination of alleles within the population. It is also thought that occasionally a new species could form if the population that remains is not geographically split; one example of this would be if a particular allele or combination of alleles made the organism choose an alternative habitat within the same area, thus effectively separating itself reproductively from the parent group.

There are a number of reasons why there can be different frequencies of alleles between two separated populations. These include *genetic drift*, which is when by chance, in a small population, the gametes (sex cells) produced at meiosis (sexual cell division) form a biased sample of allele frequencies compared with the original parent population. There is also the *founder effect*, which is when a small population, separated from the original population, has a limited variety of alleles to start with and this results in differences in allele frequencies from the parent group.

Well-known examples of organisms which formed into different species due to geographical separation include the fruit flies (*Drosophila*) on the islands of Hawaii, the Cichlid fishes of Lake Victoria (where in the past there have been numerous separate lakes at different times) and Darwin's finches on the Galapagos Islands.

All of this is widely known as *micro-evolution*. All the changes noted are small and seem more like tinkering with the original stock rather than forming anything radically new. The process is known as *recombination* (of different alleles). Few people would ever argue against the fact of micro-evolution. There are clear cases such as those mentioned and there is a clearly understood mechanism of reproductive isolation leading to different frequencies of the original alleles. Many people think, however, that this is what evolution is all about and they point to micro-evolution as if the problem is solved. They could not be more wrong. Micro-evolution is micro; tinkering and shuffling the existing genes so that varieties are produced which can eventually form closely related species. In this scenario *there is no new information in the DNA* and so the possibilities are distinctly limited.

Mutations and macro-evolution

Micro-evolutionary theory cannot explain how brand new structures and body plans can appear. Such major changes are said to occur because of what is termed *macro-evolution*. If you read a textbook on this subject, such as the one mentioned from the Open University, then you will find that there is much detail given about the analysis of alleles in micro-evolution but that for macro-evolution there is simply the bland statement that *mutation* is the driving force for change – with surprisingly little evidence given or explanation of how this can be.

Mutations occur when there is a mistake during replication of the DNA molecule and geneticists recognise two main types. In *gene mutation* an allele of a gene changes, becoming a different allele. Because such a change occurs at one gene, on one point of the chromosome, it is sometimes called a *point mutation*. In contrast to this there is *chromosome mutation*, when segments of chromosomes change. When a point mutation occurs there can be three possibilities: There is substitution of a nucleotide base, addition of a nucleotide base or deletion of a nucleotide base. You will remember that there are four DNA nucleotides A, G, C and T. A always pairs with T on the opposite chain and C with G. If a base is added during replication then one will be added on the other chain also (its complement as described above). If one is substituted then likewise there is a substitution on the opposite chain.

You will remember that the nucleotide bases code for amino acids in triplets. There are various possibilities, therefore, when a point mutation occurs. There may be a new amino acid coded for and inserted into the protein (for example, when a substitution results in a different triplet). Or the same amino acid may be coded for (because some amino acids are coded for by more than one triplet) and the mutation is therefore *silent*. Alternatively, if there is a deletion, there may be a completely new amino acid sequence for the protein coded from that point in the DNA – this is because the reading of the base sequence goes out of phase and totally different triplets may be read from it from that point on. This is known as a *frame shift* mutation and it usually results in a non- functioning polypeptide or protein. Sometimes there may be a new, or substituted, amino acid in the protein chain but this has no known effect and so the mutation is *neutral.* It appears, for example, that the haemoglobin molecule can have a certain amount of variety without losing its overall function – we know this because of studying haemoglobin molecules from different species. Another possibility is that a triplet may be formed which codes for a stop in the reading (known as a stop codon) – from there the amino acid chain construction will also stop, resulting in an incomplete protein. Sometimes larger sequences of the DNA are deleted, changed or duplicated. Large parts of the DNA molecule, known as *introns,* up until very recently, have no known function and mutations in these regions have no obvious effect. I discuss introns – so-called '*junk*' DNA below.

Chromosomal mutations involve various alterations in structure which cause harm or loss of function. Increased number of chromosomes (*polyploidy*) occurs fairly often in plants and can result in new varieties – often larger than the parent plants. It must be emphasised, though, that this does not confer any really new information in the DNA and therefore cannot be a mechanism to explain macro-evolution.

Cells have numerous sophisticated enzymes which repair mutant sites on the DNA and which prevent the vast majority of mutations being expressed. You may remember me mentioning these in the first chapter, where we looked at the constraints on large sections of DNA replicating without such repair mechanisms (as had to be for the first cells in a Darwinian model). There are a host of specific enzymes, some of which prevent damaging compounds getting to the DNA and others that cut and splice the DNA in numerous ways to remove harmful mutations.

These mechanisms go to show that mutations are generally harmful and must be prevented for organisms to survive.

As a doctor I am very aware of the devastating effect of mutations in human beings. Mutation, to a doctor, signifies damage and often leads to conditions incompatible with life. Can we be sure that mutations are the source of the genetic information that leads to useful new processes and structures in organisms? This is the key question that needs to be answered by geneticists.

Possibly the most detailed and sophisticated textbook on genetics available is *An Introduction to Genetic Analysis* by Griffiths, *etal.*[2] These are heavyweights in the genetic world. In their textbook there is a large section on the mutation of genes and chromosomes and another on evolution (a surprisingly small section). Much of what I write here is based on this book and here, if anywhere, we should see the evidence for mutations leading to macro-evolution.

We find no such thing. The only mutations that are known to be useful are the acquisition of resistance of certain organisms to external agents such as pesticides or antibiotics. There are, in fact, no recorded instances of the type of macro-evolution that I have discussed above. I will quote from the book:

'Because mutation events introduce random genetic changes, most of the time they result in loss of function. The mutation events are like bullets being fired at a complex machine; most of the time they will inactivate it. However it is conceivable that in rare cases a bullet will strike the machine in such a way that it produces some new function.'[3]

I do not think that I am the only one who could *not* imagine any complex machine being improved by firing a bullet at it. In fact the machinery inside a cell is far more complex than any machine ever constructed.

There have been endless attempts to observe mutations over the last half-century. Much of the work done is on fruit flies (*Drosophila*). They are useful because they are easily bred in captivity, there are a multitude of species and they have rapid breeding cycles. Scientists have repeatedly used methods to greatly increase the mutation rate of fruit flies so that vastly more mutations can be observed. This involves subjecting them to various forms of radiation and also using chemicals (mutagens) that increase mutation rates. The result of this is that we ought to see in the laboratory some direct evidence of mutations causing new functions and forms which could conceivably be useful in the wild. No such mutations have ever been seen. The sorts of mutations that we see are usually harmful, such as stunted wings, deterioration in vision and non-functional alteration in limb position.

If we are to accept the Darwinian model then we must test its integrity. As mentioned before in discussing the work of Karl Popper, we need to predict from the theory certain expected results which it will fail if the theory is false. Certainly one prediction is that we should see beneficial mutations on the macro level after decades of subjecting billions of fruit flies to mutagens. On this test the theory fails completely.

The geneticist Steve Jones has written an updated (I have to say less well written) version of Darwin's *Origin of Species – Almost Like a Whale*[1]. In his first chapter he enjoys a good attack on creationists. (It is interesting that creationists are nearly always lumped together despite having a great diversity of viewpoints.) He points to the one bit of evolution that we have all seen in our lifetime – that of the AIDS virus HIV. He rightly explains how the virus has, through mutation, changed so that there are varieties. This is the only evolution that he can tell us about that has been observed occurring because of mutation. This deserves comment because it actually reflects the very poor evidence for major change being able to occur through mutation. The AIDS virus, firstly, is a virus – an entity which even virologists cannot say for certain is 'life'. It is not a cell and is basically a package of RNA that gets into human cells, takes over and causes vast numbers of itself to be replicated. The destruction of cells in the immune system leads to AIDS. Any changes that have been observed through mutation are very small – essentially the virus remains HIV. To parade this as evidence for mutation leading from *Pikaia* (that chordate seen in the Cambrian explosion) to you and me is simply wrong and merely shows up the paucity of evidence that there is for macro-evolution through mutation.

We need to realise what happens when a mutation occurs. There may indeed be some new information in the DNA which could mean that a different amino acid, or series of amino acids, was inserted into a protein. In another chapter we will look in more detail at the complexity of some of the intra-cellular machinery but suffice it to say now that the enzyme systems and workings of a cell are the most complex physical structures in the known universe. When a mutation causes a change to such a system then it needs to confer a definite advantage if it is going to survive into future generations. The systems in the cell are so finely tuned that alterations such as those from mutations are either harmful or make very little if any difference.

Try to imagine converting a steam-driven car into a modern petrol-driven one. The changes needed are totally radical and cannot be achieved in tiny steps. To become a petrol engine requires many changes all at once. A mutation is equivalent to putting a different fuel (petrol) into an engine designed to run on steam. It will not work. A totally new engine must be designed from scratch. The systems in a cell are far more complex and they govern the whole of the organism. What we see in the fossil record are many sudden changes which are equivalent to the conversion of a steam engine to a petrol one. We have discussed many of these and you will remember the new body plans of the Cambrian period, the appearance of fish, the appearance of tetrapods, feathers, warm blooded mammals, brains. All of these are macro changes which are not explained on mutation theory.

Another problem for the Darwinist is that mutation rates in organisms are very very slow. Even the billions of years that have elapsed are simply not adequate to achieve enough progressive genetic change to bring about evolution. Griffiths, *et al.*, in their book list some typical mutation rates for point mutations in various genes of bacteria, corn and fruit flies. Mutation rates per generation range from from 2 x 10 to the power -8, to 2 x 10 to the power -4. Griffiths, *et al.,* conclude, and I quote:

> 'Mutation rates are so low that mutation alone cannot account for the rapid evolution of populations and species.'[5]

This is a plain admission that we do not have any idea what has caused macro-evolution.

Another problem is that most mutations, if passed on to the next generation, are recessive and so the allele will not become frequent in that population. In fact there seems to be no clear way that, even if the allele is dominant, the mutant organisms can so succeed that the allele becomes the norm. The first mutant organism must be still be able to breed with the original stock in order to have progeny. No one seems to have a clear idea how any major mutation can thus become dominant in the population.

Richard Dawkins has written a number of extremely popular books which try to explain the power of evolution and the logic of how it all happened. He may be sincere but he frequently makes serious mistakes in his reasoning. In his book *The Blind Watchmaker*,[6] he tackles the whole problem of how small changes in the amino acid sequence of a protein could lead to a molecule such as haemoglobin – which carries oxygen in our blood. As he says, the haemoglobin molecule consists of four chains of amino acids twisted together. Each chain consists of 146 amino acids. He attempts to describe how such a complex protein could have come about by small, chance steps – using a computer model. I describe this in detail in my chapter on his book *The Blind Watchmaker*. There I will show that his model is flawed and that there is no logical way that we can see an irreducibly complex molecule such as haemoglobin appearing through random mutations – no matter how long the time-frame is.

I discuss the probability of random mutations leading to something as complex as the human eye in Chapter 9 and will show that it is so unlikely that 'miracle' can be the only appropriate term for such a sequence of events.

Molecular Phylogeny

A relatively new discipline related to all this is that of *molecular phylogeny*. It is now possible to work out the sequences of proteins, RNA and DNA in different organisms. By studying these it is hoped that evolutionary relationships might be unravelled among the varieties of life. For example, ribosomal DNA is studied to try to build up a picture of the relationship between the 30 or so existing phyla or body plans of animals. Evolutionary theory predicts that small changes over time in such sequences have led to specific sequence patterns which allow us to classify the various groups in relationship with one another. The more similar are the molecules then (the theory predicts) the closer are the organisms on an evolutionary tree. We touched on this in the first chapter when discussing the root of the so-called 'tree of life'. Also, there is the idea that mutations tend to occur at a steady rate and so act as a sort of clock – the more the differences there are in a molecule (of say haemoglobin) between two species, the further back is the time when the two species supposedly diverged from a common ancestor.

A good overview of this subject is given by Roger Lewin[7]. He shows us that there is actually no consensus from this work about the relationships of the phyla. Nor is there any clear correlation between relationships based on morphology (the physical characteristics of organisms) and molecular phylogeny. It is also very unclear whether the molecular clock idea is valid or accurate. Many doubt that the clock is running at a steady rate. There is a good deal of circular thinking amongst the experts – not least in trying to date the appearance of humans in the world. Everything depends on your prior view of whether, for instance, we descended from a common ancestor shared with apes or not and whether the molecular clock is an accurate tool to date evolution. Our relationship to apes is discussed in Chapter on 6.

'Junk' DNA

Enormous sections of DNA have sequences without apparent function. These so-called 'non-coding' sections, or introns, occupy an astonishing 98 per cent of human DNA. When we speak of the human genome being mapped, what is spoken of is the 2 per cent of DNA that constitute genes. To evolutionists, such as Richard Dawkins, these enormous sections of apparently useless DNA are evidence for non-purposeful evolution and are just remnants of previously used genetic material which has been superceded (he goes into some detail about this is his book *The Blind Watchmaker*). He likens this 'junk' DNA to the sections of binary information on a computer disk that are still present after you hit the delete button.

My reaction to this, as a Christian who believes in purpose behind all life, has always been to assume that one day we will find that this junk DNA actually does have function. Recent work has confirmed this and has shown that, contrary to previously held dogma, some of these areas of DNA contain coding for certain types of RNA which are very active in regulating the cell. Some of these very short RNA sequences act in regulating genes and have profound influence over how genes are expressed. This long neglected part of DNA has only very recently been investigated and there is a feeling in the air that genetics is going to have another revolution.[8]

This discovery of the uses of intron DNA may help clear up a puzzle: there is no clear correspondence between the complexity of a species and the number of genes in its genome. For example, fruit flies have fewer coding genes than roundworms and rice plants have more than humans. The present estimate is that the human genome contains 27,000 genes. If intron DNA has profound effects on organisms, then this may explain why numbers of conventional genes are by no means the whole picture.

Homeobox (Hox) genes

Among the most fascinating developments in genetics of the past decade is the discovery of *homeobox* genes, some of which are known as *Homeotic* genes. These are genes which co-ordinate some of the most basic processes of the early development of animals. They determine such arrangements as the front-to-back and top-to-bottom orientation of organisms and the order and position in which certain organs and structures appear.

There are some surprising and unexplained facts about these genes. For example, in some cases genes that occur at the front end of the chromosome (known as the 3' end) have their effect mainly at the front end of the animal, whereas those at the back end of the chromosome concerned (known as the 5' end) have their effects more posteriorly.

More amazingly, some of these genes do similar jobs in very different animals. The *Hox6* gene group, for instance, which gives the instruction 'make legs' is much the same in fruit flies and tetrapods. Similarly, a gene known as *Pax-6* (a homeotic gene that is not technically in the *Hox* group) is involved with the development of the camera eyes of vertebrates, as well as the compound eyes of arthropods. This is extraordinary. I quote from molecular biologist Jonathan Wells:

'since Pax-6 affects such radically different structures as compound (multifaceted) insect and camera-like vertebrate eyes, some biologists have suggested that its original function was to specify an ancestral light-sensitive spot. But fruit flies possess simple eyes as well as compound eyes, and their simple eyes (the closest thing they have to light-sensitive spots) are unaffected by Pax-6; ... the universality of homeotic genes is supposed to be due to their presence in a common ancestor, but the preponderance of evidence suggests that the common ancestor lacked the features that those homeotic genes now supposedly control. From a Darwinian perspective this is a serious problem. According to neo-Darwinism, complex gene sequences gradually evolve by conferring selective advantages on the organisms that possess them. But the gene sequences confer selective advantages only if they program the development of useful adaptations. If a primitive animal possessed homeotic genes but lacked all of the adaptations now associated with them, then those genes must have originated prior to those adaptations. How then did homeotic genes evolve?'[9]

All this points to some deep genetic congruence between these widely different animals. The Darwinist will say it points somehow to a common genetic past – but, as Jonathan Wells shows us, this is hard to reconcile when the organs concerned have appeared quite separately in the history of life (for example, the compound eyes of arthropods/the camera eyes of vertebrates; the legs of fruit flies/The legs of tigers). Far more likely is the presence of a common designer.

The study of genetics has opened up to us the mechanisms of

inheritance and Darwinists have seen this as verification of the theory of evolution. We have seen the validity of this as far as micro-evolution is concerned but for the large-scale changes in organisms which we see in the fossil record, macro-evolution, there is no known mechanism. Mutation theory completely fails to account for such changes.

1 Peter Skelton (1996). Evolution. A biological and palaeontological approach. Addison Wesley in association with Open University.

2. Griffiths, Miller, Suzuki, Lewontin and Gilbart (1999). *An Introduction to Genetic Analysis*. W. H. Freeman.

3. Ibid., p472.

4. Steve Jones, (1999). *Almost Like a Whale*. Anchor.

5. Griffiths, *etal.*, *An Introduction to Genetic Analysis*, p724.

6. Richard Dawkins (1986). *The Blind Watchmaker*. Longman Scientific and Technical.

7. Roger Lewin (1996). *Patterns in Evolution. The New Molecular View. Scientific American*8. See W. Wayt Gibbs (2003) The Unseen Genome: Gems along the Junk,.

8. See in W. Wayt Gibbs. The Unseen Genome: Gems among the Junk. *Scientific American.* Nov. 2003.

9. Jonathan Wells (1998). Unseating Naturalism. In *Mere Creation*. Ed. William Dembski. Intervarsity Press.

Chapter Eight

Irreducible Complexity

In 1996 Michael Behe, Professor of Biochemistry at Lehigh University, wrote a book that may well prove to be historic in the turning of the tide against Darwinism.[1] In his book he elegantly describes many of the complex biochemical systems that are found in living organisms and he pronounced them *irreducibly complex.*

He uses the simple mouse trap to explain the concept. A mousetrap (of the old-fashioned sort) consists of a few parts that together make an effective trap. They are the platform, the spring, the hammer (bar which comes down on mouse), the holding bar and the catch (see diagram in figure 8.1).

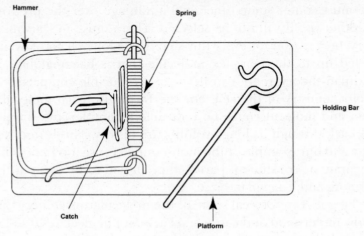

figure 8.1

Each of these parts is essential to the proper functioning of the trap. However, it is quite obvious that any one of the parts is useless on its own. The spring, for example, can have no function if alone. The trap is irreducibly complex in that you cannot reduce it to its constituent bits– it must be all together or not work. Opponents of Behe have tried hard to show that a mousetrap can actually catch mice or be used effectively in its separate parts. I feel, reading such arguments, that they are rather desperate attempts to hold back the tide.

In Darwinian evolutionary theory, all the molecules and structures that make up an organism must have evolved from something else. Every molecule, such as haemoglobin, must have come about through a series of short steps from some ancient precursor. The odds of getting a fully formed haemoglobin molecule from scratch are astronomical, as Richard Dawkins has described. If the Darwinist could just see the whole history of that molecule he would hope to find a very simple start which nature threw up randomly, followed by a large number of intermediate stages gradually leading up to the present-day molecule.

It is vitally important to grasp that, in Darwinian theory, each stage must confer some functional advantage over the previous one, otherwise it will not be selected. There must, in Darwinian theory, be a reason for every tiny step in the long chain of events that led up to the complex molecule that is haemoglobin. In Darwinian theory, there can never be irreducible complexity in the biochemistry of the cell. For the Darwinist, each molecular system and molecule must be reducible to simpler and simpler parts back through its long evolutionary history. This is in direct contrast to our example of the mousetrap because it is impossible to separate its constituent parts – they all had to be there at the beginning and *designed* to be so.

Behe goes on to reveal some of the molecular machinery that we now understand and can marvel at – and he declares it to be, just like the mousetrap, irreducibly complex. There can be no way, he says, that these systems could have evolved. They had to be fully functionally complex as they are now and could not have arisen from simpler beginnings. In short, they were designed.

Design is an embarrassing word to most biologists these days. This is in total contrast to the way that biologists thought before Darwin's theory appeared. Design means that some intelligent external being has purposed and brought about the systems that make up organisms. This is the antithesis of the idea of the essentially blind and random mutations that are key to Darwinism.

In this chapter I am going to give just two examples of irreducible complexity which Behe describes.

The Bacterial Flagellum

It is easy to assume that the mechanisms in the lowly bacterium are somehow less sophisticated than in so-called higher animals. You would be wrong. Some bacteria swim with the aid of a flagellum, which is a structure not found in more complex cells. It acts as a sort of rotary propeller. The structure of the flagellum is quite different from that of a cilium. It is a long, hairline filament which consists of a protein called 'flagellin' and this is embedded in the cell membrane (see figure 8.2).

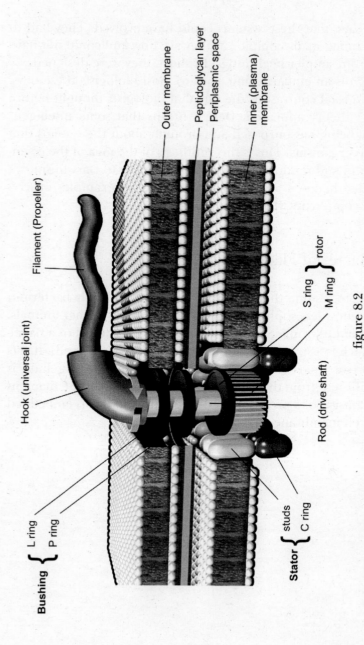

figure 8.2

Drawing of a bacterial flagellum showing the filament hook, and the motor imbedded in the inner and outer cell membranes and the cell wall.

I will quote Behe at this point:

> 'The flagellin filament is the paddle surface that contacts
> the liquid during swimming. At the end of the flagellin fil-
> ament near the surface of the cell, there is a bulge in the
> thickness of the flagellum. It is here that the filament
> attaches to the rotor drive. The attachment material is com-
> prised of something called 'hook protein'. The filament of
> a bacterial flagellum, unlike a cilium, contains no motor
> protein; if it is broken off the filament just floats stiffly in
> the water. Therefore the motor that rotates the filament-
> propeller must be located somewhere else. Experiments
> have demonstrated that it is located at the base of the fla-
> gellum, where electron microscopy shows several ring struc-
> tures occur. The rotary nature of the flagellum has clear,
> unavoidable consequences, as noted in a popular biochem-
> istry textbook: 'The bacterial rotary motor must have the
> same mechanical elements as other rotary devices: a rotor
> (the rotating element) and a stator (the stationary ele-
> ment)'. The rotor has been identified as the M ring in fig-
> ure 8.2 and the stator as the S ring.'[2]

Behe goes on to explain that, unlike other systems that gener-
ate motion (such as muscles) the bacterial motor does not directly use energy that is stored in a 'carrier' molecule such as ATP.
Instead it uses energy generated by a flow of acid through the bacterial membrane. This is complex and is the subject of active
research. A number of models are proposed and figure 8.3 gives
a taste of the complexity.

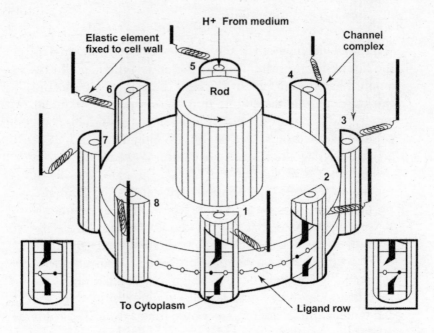

figure 8.3

The flagellum uses three main parts: a paddle, a rotor, and a motor – as such it is irreducibly complex. To evolve gradually is logically impossible for such a system and there is no literature that shows any model to explain such a gradual evolution.

It is even more complex than at first glance; the flagellum requires about 40 other proteins for function. The roles of these include signals to turn the motor on and off; 'bushing' proteins to allow the flagellum to penetrate through the cell wall; proteins to assist in the assembly of the structure; and proteins to regulate the production of the proteins that make up the flagellum. It seems that the more is discovered, the more complex it gets – and new knowledge, far from alleviating Darwinian problems, only serves to make them more intractable. The bacterial flagellum is a good example of an irreducibly complex structure in life.

The biochemistry of vision

Darwin had no idea how vision was accomplished. Often he pondered over the complexity of the eye but knew nothing about the amazing job that molecules do in the retina. The following is a summary of this biochemistry – as such it will be a hard read for many. As Behe says, do not be put off by the names of molecules; they are only labels and no more esoteric than *carburettor* or *differential*. I will quote the next few paragraphs verbatim from Behe. It is written in italics to show that you can skip it if you wish. But it is better to have a go at skim reading to get an idea of the irreducible complexity.

Figure 8.4 shows the change of shape of rhodopsin after a photon hits it.

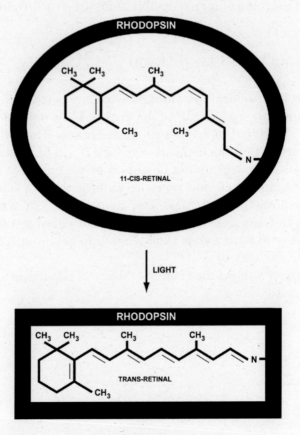

figure 8.4

The first step in vision. A photon of light causes a change in the shape of a small organic molecule, retinal. This forces a change in the shape of the much larger protein, rhodopsin to which it is attached. The diagram above is not to scale.

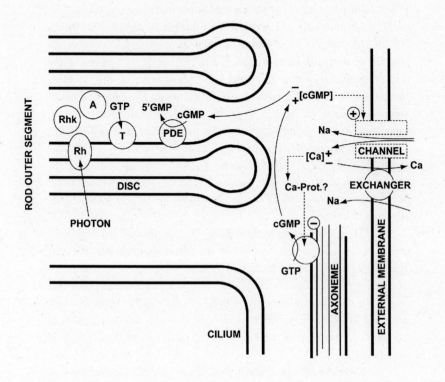

figure 8.5
The biochemistry of vision. RH, Rhodopsin; RHK, Rhodopsin Kinase; A, Arrestin; GC, guanylate cyclase; T, Transducin; PDE, Phosphodiesterase

Now for some technical information:

When light first strikes the retina a photon interacts with a molecule called 11-cis-retinal, which rearranges within picoseconds to trans-retinal. (A picosecond is about the time it takes light to travel the breadth of a single human hair.) The change in the shape of the retinal molecule forces a change in the shape of the protein, rhodopsin, to which the retinal is tightly bound. The protein's metamorphosis alters its behaviour. Now called metarhodopsin II, the protein sticks to another protein, called transducin. Before bumping into metarhodopsin II, transducin had tightly bound a small molecule called GDP. But when transducin interacts with metarhodopsin II, the GDP falls off, and a molecule called GTP binds to transducin. (GTP is closely related to, but critically different from, GDP.)

GTP-transducin-metarhodopsin II now binds to a protein called phosphodiesterase, located in the inner membrane of the cell. When attached to metarhodopsin II and its entourage, the phosphodiesterase acquires the chemical ability to 'cut' a molecule called cGMP (a chemical relative of both GDP and GTP). Initially there are a lot of cGMP molecules in the cell, but the phosphodiesterase lowers its concentration, just as a pulled plug lowers the water level in a bathtub.

Another membrane protein that binds cGMP is called an ion channel. It acts as a gateway that regulates the number of sodium ions in the cell. Normally the ion channel allows sodium ions to flow into the cell, while a separate protein actively pumps them out again. The dual action of the ion channel and pump keeps the level of sodium ions in the cell within a narrow range. When the amount of cGMP is reduced because of cleavage by the phosphodiesterase, the ion channel closes, causing the cellular concentration of positively charged sodium ions to be reduced. This causes an imbalance of charge across the cell membrane that, finally, causes a current to be transmitted down the optic nerve to the brain. The result, when interpreted by the brain, is vision.

If the reactions mentioned above were the only ones that operated in the cell, the. supply of II-cis-retinal, cGMP, and sodium ions would quickly be depleted. Something has to turn off the proteins that were turned on and restore the cell to its original state. Several mechanisms do this. First, in the dark the ion channel (in addition to sodium ions) also lets calcium ions into the cell. The calcium is pumped back out by a different protein so that a constant calcium concentration is maintained. When cGMP levels fall, shutting down the ion channel, calcium ion concentration decreases, too. The phosphodiesterase enzyme, which destroys cGMP, slows down at lower calcium concentration. Second, a protein called guanylate cyclase begins to resynthesize cGMP when calcium levels start to fall. Third, while all of this is going on, metarhodopsin II is chemically modified by an enzyme called rhodopsin kinase. The modified rhodopsin then binds to a protein known as arrestin, which prevents the rhodopsin from activating more transducin. So the cell contains mechanisms to limit the amplified signal started by a single photon.

Trans-retinal eventually falls off of rhodopsin and must be reconverted to II-cis-retinal and again bound by rhodopsin to get back to the starting point for another visual cycle. To accomplish this, trans-retinal is first chemically modified by an enzyme to trans-retinol – a form containing two more hydrogen atoms. A second enzyme then converts the molecule to II-cis-retinol. Finally, a third enzyme removes the previously added hydrogen atoms to form II-cis-retinal, a cycle is complete.[3]

This is the bare bones of the biochemistry of vision. It is more complicated. Behe explains that this system cannot be reduced to its component parts; all of the intricate biochemistry involved must be there complete – or not at all.

Darwin, and evolutionists since, have looked at the gross anatomy of the eye only, and have generally failed to tackle the nuts and bolts of the molecular system of vision. As such they have avoided real areas of difficulty for the theory. The biochemistry of vision is irreducibly complex. There is no way that the pathways and molecules could have achieved their intricate and, dare one say, beautiful, system by small steps. Each part of the whole is superbly fitted to the system and cannot work in any independence of it. Like the humble mousetrap, the visual system must have been designed.

Behe gives more examples from his speciality – such as the blood clotting system. In reality, there are hundreds, if not thousands of other irreducible biochemical systems that we now know of. I will deal with the irreducible complexity of haemoglobin, the oxygen carrying molecule in our blood, in Chapter 13. This will show up some of the flawed arguments of Richard Dawkins – who has attempted to show that haemoglobin could have come about in small steps.

If it is really true that the systems within cells are not reducible to simpler parts then this is a clear argument for design and a complete refutation of Darwinism. I would encourage you to read Behe's book.

1. Michael Behe (1996). *Darwin's Black Box*. The Free Press
2. Ibid., P.70
3. Ibid., pp.18-21

Chapter Nine

The Impotence of Natural Selection

Darwin's theory depends on selection of the fittest individuals and a steady amount of variation within the population to provide variants upon which selection works. It was vital for his theory that this variation should be copious, in small steps and to be non-directional.

It must be copious because there needed to be a fund of variety to account for the huge disparity in forms of life and their complexity. It must be in small steps because, to him (and modern Darwinians) a theory that depends on large-scale leaps of form (known as saltations) must require some internal driving force within organisms to produce such forms. Darwin could not stomach this idea because it smacks of teleology – purpose and design. If there are sudden macro leaps in new organs or anatomy then there has to be some other force which provides this apart from selection. Selection cannot provide sudden large-scale new forms. Darwinism completely depends on there being only physical causes for evolution, so any internal *direction* within a lineage of organisms, which saltation implies, is unacceptable.

The variations must be non-directional because to speak of *direction* implies non- scientific ideas of purpose and mystery. It was important therefore to Darwin that variants could go in any direction, and certainly not just advantageously.

Many believe, quite wrongly, that ever since Darwin there has been a consensus amongst biologists that there is no internal drive to evolve and that the variants, which provide fodder for selection to work upon, are tiny and gradual. However, the truth is that until the 1930s there were many amongst the leading figures of the biological establishment, who believed in either saltations or an internal drive or direction to the production of new forms. After the publication of Darwin's *Origin of Species* Thomas Huxley, Darwin's vocal defender, was less than convinced that Darwin needed to emphasise the tiny nature of the variations needed for evolution. He, and many others, felt that the immense scale of variety within the animal kingdom points to the need for macro leaps in development of organisms. Charles Lyell, Darwin's friend, whose work in geology paved the way for Darwin's theory, also felt unconvinced that natural selection on tiny variants was enough to produce the abundance of life in our world.

Darwin had a brilliant and eccentric cousin called Francis Galton – a leading theorist about evolution of the late nineteenth century. He argued that continuous small variations could not produce major changes. He believed that in any population of organisms, such small variations would show *regression towards the mean* – the variations would be averaged out within the population, so no major advance could occur. He then proposed a theory of discontinuous variation whereby evolutionary advances could only occur in saltations. To illustrate this he used the metaphor of a polyhedron; a sort of polygonol slab (see figure 9.1).

The organism is likened to this polyhedron which rests on a flat surface. Advance can only be made by rolling the slab so that a new facet of the polyhedron rests on the flat surface. The normal state is one of stasis but if sufficient push is given then the slab will roll. Sometimes it will teeter on a corner only to fall back into the same place (as occurs with small variations) but occasionally a large variation occurs so that it will flip to a new facet. Interestingly, this idea does fit with what we see in the fossil record; long periods of stasis followed by sudden changes.

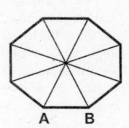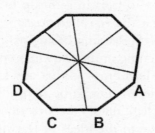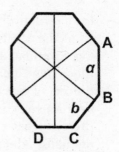

figure 9.1

Galton's own illustration of his model of the polyhedron. Note how the themes of saltation, or facet flipping, and constraint in strictly limited pathways available for change arise from a similar geometric basis in this mode of depiction.

This image of facet flipping polyhedrons provided ideas for some of the major evolutionary thinkers of the following decades. For a full (and excellent) account of the history of these ideas then essential reading is *The Structure of Evolutionary Theory by Stephen Jay Gould*[1].

One major idea that was entertained is that of *orthogenesis*. Orthogenesis (a term coined by the German zoologist Wilhelm Haacke) denotes the idea that evolution proceeds along defined and restricted pathways because internal factors limit and bias variation into specified channels. As such this theory stands against that of Darwin, whose principle was that only natural selection imparts direction by shaping isotropic (non-directional) variation. This Darwinian principle, already outlined above, requires that selection is not simply acting in a negative way that eliminates the unfit while some other process creates the fit. Leading naturalists such as Theodor Eimer, Alpheus Hyatt and C. O. Whitman in different ways argued for orthogenesis.

The second major group of ideas for an internal process that drives evolution (leaving Darwin's natural selection as subsidiary) is that of already mentioned saltations. The leading figures in this movement were William Bateson and Hugo de Vriess.

Both orthogenesis and saltation were very much on the agenda for alternative views to Darwin right up to the 1930s. It was only from the 1930s that the so-called *modern synthesis* of Darwinism was formulated by the leading biologists of that time. Bit by bit there came a hardening of the Darwinian paradigm by the evolutionists of our modern era – so that any talk of orthogenesis or saltations is considered now to be beyond the pale. There is at present what can legitimately be described as a sort of fundamentalist adherence to the ideas of Darwin. It is only some few thinkers such as the late Stephen J.Gould who are prepared to put their heads above the parapet and question some of the rigidly held tenets. Gould was, of course, no creationist but he did reformulate some of the errors of this hardened synthesis – in particular he resurrected the idea of internal evolutionary direction through predefined channels within the species and also the idea that species (and not just individual organisms) can be selected upon. He also maintained that not every aspect of an organism is necessarily chosen by natural selection for a specific adaptation – but may be simply a side effect of another adaptation.

This brief historical introduction opens up the way to discuss the whole problem of where the creativity comes from for the production of new forms of life. The Darwinian, as already explained, holds strenuously to the view that it is only the external pressure of selection acting on copious, non-directional and small variations that results in new species and new body plans. I believe, however, that the doubts of the major players since Darwin (up to the 1930s) were well founded. Most of them were not creationists – in the sense that they felt the processes to be entirely under physical laws. However, their doubts about Darwin's theory badly need to be looked at afresh so that we can find logical answers where they are distinctly lacking in fundamental Darwinism. Their doubts, often brilliantly expounded, have been conveniently forgotten and few now read the works of, for instance, Hugo de Vriess – a first-class botanist who probably

understood Darwinism better than anyone else at the beginning
of the twentieth century. By studying these scientists we learn that
selection is not enough to explain the appearance of new organisms. The following is a quote from Bateson:

> 'To begin with, we must relegate selection to its proper
> place. Selection permits the viable to continue and decides
> that the nonviable shall perish; just as the temperature of
> our atmosphere decides that no liquid carbon shall be
> found on the face of the earth: but we do not suppose that
> the form of the diamond has been gradually achieved by a
> process of selection. So again, as the course of descent
> branches in the successive generations, selection deter-
> mines along which branch evolution shall proceed, but it
> does not decide what novelties that branch shall bring
> forth.'[2]

And here is De Vriess:

> 'Natural selection is a sieve. It creates nothing, as is so often
> assumed; it only sifts. It retains only what variability puts
> into the sieve. Whence the material comes that is put into
> it, should be kept separate from the theory of its selection.
> How the struggle for existence sifts is one question; how
> that which is sifted arose is another.'[3]

See also the work of Michael Ruse for a history of ideas concerned with Darwinism.[4]

Selection has been used by Darwinians to explain the emergence of the immense variety of life on earth. But what is selection? Selection, as so clearly stated by De Vriess, is like a sieve. As
a sieve it allows some organisms to go through (and die) while a
few are retained and survive. This is all that selection can actually do. Selection is not creative. This needs emphasising: *selection is*

not in any way creative. All it can do is weed out some and allow others to survive. In many ways it is therefore a negative force only – denying some organisms life.

In Chapter 13 I will discuss the problems we have in understanding the evolution of the eye – as a critique of Richard Dawkins who has tried to explain it. Dawkins admits that the eye is incredibly complex and he attempts to show how it could have arrived in stages through natural selection acting on each randomly produced novelty. He admits that the chances of an eye appearing by chance in one go are vanishingly small. He makes the mistake though of assuming that each tiny step-like change in his scenario of eye evolution is a credible part of a long and not impossible process. His key argument, as with all Darwinists, is that selection works on the small changes and accumulates these small improvements to produce something as complex as an eye. Where he and others falls down is in not properly understanding what De Vriess knew – that every tiny random novelty that is thrown up has to arrive *prior* to any selection acting on it. In the case of the eye, for example, there would need to be a series of very unlikely steps needed to form the pupil in front of the lens. The pupil is a diaphragm which alters the amount of light coming into the eye in response to complex nerve feedback mechanisms. To see it arriving in one go is clearly impossible – so Dawkins and others must try to imagine it coming in stages. There is no fleshing out of these stages, by the way, in Dawkins' account – it is just assumed. Where Darwinist logic falls flat is in trying to assess probabilities. As said, each new novelty required in the stages of pupil evolution must arrive by chance prior to being selected. (I cannot emphasise more the need to see each new randomly produced change in evolution as something that would have to come about *prior* to selection acting on it.) Another mistake which many, including Dawkins make is seeing the eye as a unit separate from the brain. The retina is, in fact, an extension of the brain and we must view eye evolution as going hand-in-hand with enormously complex changes in brain anatomy and physiology – not only in the visual cortex but in many other neural connections.

A key concept, which is also confused by Dawkins and others, is the idea of his that once a primitive eye is formed then all that is needed is quantitative changes in what is there already (a little bit more or less of the parts of the primitive eye). This idea helps him to manage the probabilities. In real life though it is the *qualitative* changes that are improbable. All along the supposed evolution of the eye there would need to be brand new structures and systems put in place. It is worth viewing the anatomy of the eye from the external muscles inwards, right down to the irreducibly complex visual biochemistry and also the myriad of brain connections. All is novelty. To see the evolution of the eye as mere tinkering with an ancient 'proto-eye' is to live in cloud cuckoo land.

No one knows what the actual probabilities of the mutations required for eye evolution are. However, the maths is not too difficult: if there are 10,000 steps needed in evolving an eye (and there would have to be many more of course), then each new mutational step must arrive by chance with a probability of it arriving. By multiplying such probabilities we can arrive at some sort of figure of how likely it is to happen.

If each mutation has a probability of occurring in one in 10 million individuals (I would see this as a very conservative estimate), then the chance of getting all the 10,000 mutations in series is 10 million times itself 10,000 times. We arrive at the probability of a series of such mutations occurring to be one chance in 10 to the power 70,000. This last figure greatly exceeds the number of atoms in the universe.

A Darwinist may then accept that the human eye is a highly improbable structure but may argue that it is just one of many other types of potential eye that evolution could throw up. Perhaps we just happen to have this visual system which is one of many alternative ones. This idea helps the evolutionist to scale down the improbability of some such eye occurring. This might just be a valid argument were it not for the almost identical eye of the octopus. We know that the octopus eye is like our own – uncannily so. Yet we also know that the octopus comes from a totally separate phylum and would have had a completely separate evolutionary history. To have two separate cases of such an improbable series of events is simply beyond credence.

We have seen that natural selection in itself cannot provide novelty and merely selects what variation is there. This clearly has an effect – we are not saying that natural selection does nothing at all. We see the effects of natural selection in choosing the fittest to survive. We see its effect in favouring the better camouflaged, the faster, the stronger etc. What it cannot do, however, is create those variations that it selects. De Vriess' idea of it being a sieve is very helpful. A sieve is inert and uncreative.

If selection is not creative then where does the abundance and complexity of life come from? This question will not go away. There are no mechanisms that we can see or even imagine in the genetics of life that will produce a creature such as a human.

We are left with no alternative but to see another force at work – one which modern science does not feel comfortable with but with which we must reckon. We are made by design and purpose and with a power that we have no idea of. Let us rescue the real meaning of the word *creature*. Creatures are created.

1 Stephen Jay Gould (2002). *The Structure of Evolutionary Theory*. Harvard: Belknap Press.

2. W.Bateson (1909). 'Heredity and Variation in Modern Lights', in *Darwin and Modern Science*. Cambridge University Press.

3. H. De Vriess (1909). *The Mutation Theory*. Appleton & Co.

4. Michael Ruse (1996). *The Darwinian Revolution. Science Red in Tooth and Claw*. Chicago: University. of Chicago Press.

Chapter Ten

Human Nature and Darwinism

The theory of evolution must explain all aspects of life if it is to stand. It must not just account for our physical state or anatomy but must also explain the human mind. I believe that when we look at certain aspects of the mind we have no alternative but to reject the Darwinian explanation.

Altruism

Altruism is manifest when a human being behaves in a way that is entirely for the benefit of another (or others) and is at the expense of himself. Altruism is a characteristic of human behaviour that has received much study by evolutionary writers and psychologists. This is because it does not fit easily into the Darwinian paradigm. It goes against the flow and seems to fly in the face of the ruthless competitive world of natural selection.

Let us establish first the reason why it is a problem. If it is true that we are here due to many millions of years of competition through natural selection then we should expect certain characteristics of human behaviour. Darwinism maintains that everything in our make-up has been honed for survival. Our genes have been selected purely on the grounds of how well they continue to replicate down the generations. In this law of the jungle there can be no sentimentality. We are talking about a heartless drive to succeed in passing on our genes. Nothing else, in

Darwinian terms, really matters. And so our mental make-up too must fit into this framework. We should expect a ruthlessness, however disguised. We should expect that our minds are the way they are purely because of survival. There should, in fact, be nothing genuinely altruistic about us. By genuinely altruistic I mean a form of goodness that is a free choice – not simply learnt or instinctive but chosen. Such genuine altruism implies that there is such a thing as free-will. Often real altruism is a decision to go the extra mile for someone – not just a warm feeling. If such genuine altruism exists or is even common in humans, then most would agree that there is no evolutionary explanation for it.

The evolutionists recognise this fact and have made valiant attempts to rationalise altruism and to downgrade it. It cannot, in evolutionary terms, be genuine. Any apparent altruism must be for the furtherance of our own genes and not actually selfless at all. Such apparent selflessness is just window-dressing for an underlying drive to replicate our genes.

Few people recognise how serious a problem this is. Forgive me for repeating myself here. If we are to accept the whole Darwinian package then we must also accept that we are at the end of a very long lineage which has survived only because of intense competition. Every adaptation, including our minds, must have evolved purely to increase the chance to reproduce successfully. Any diversion from this would have been wiped out very quickly and any idea that there is genuine good or love must be false. This is because genuine goodness and love is selfless and does not seek its own benefit nor does it confine its selflessness to our immediate relations. True love of one's neighbour involves self-sacrifice and this would be totally maladaptive as far as evolution is concerned.

Not very long ago an airliner crashed into a frozen river and it began to sink, full of passengers. I remember seeing the harrowing pictures on TV. Many managed to get out and were swimming around in the freezing water or clinging to the plane as it sank. A helicopter began to rescue people, hauling them one by one out of danger. One man, who was very strong, deliberately stayed behind to help other people to safety. He repeatedly swam to others, brought them to the winchman and saw them go up safely. After saving many people he was exhausted and could not get to save any more. He sank and drowned.

This man is a hero – rightly so. Yet, in Darwinian terms he was a total failure. His efforts to save those around him, who were unrelated to him, were maladaptive. He failed to pass on his own genes because he paid the ultimate sacrifice. Such behaviour could only be weeded out in the entirely ruthless world of natural selection. Such behaviour would not be selected for. Yet, he is our hero. Why, if we are the product of entirely impassive selection, do we revere such people? To revere such sacrifice is itself maladaptive because it would encourage you or me to do the same type of thing – sacrifice ourselves for a greater good.

The philosopher Anthony O'Hear has written a helpful book about this type of issue. In *Beyond Evolution. Human nature and the limits of evolutionary explanation*,[1] he brings to our attention the case of Socrates, who felt it was 'best' to stay behind in Athens and to die. Die he did, to uphold what he thought was good and right. Ever since then Socrates has been an inspirational figure someone who stands for doing the right thing no matter what the personal cost. O'Hear says that evolutionary theory cannot explain this and he is right.

To the Darwinist, apparently selfless behaviour can only be programmed into our make-up for the purpose of preserving our genes. The main way out of the problem is to assume that any inbuilt tendency to be altruistic is a means of helping the immediate community or family of the individual concerned. This is particularly seen if a person might die to save a group of his near relatives. It is not too difficult to calculate that if that person dies to save several close relatives, then, overall, he has preserved more of his genes than if they had died and he had survived. This is because close relatives share a large number of our genes.

Darwin spent some time in his book *The Descent of Man and Selection in Relation to Sex*, discussing man's moral sense, virtuous behaviour and conscience.[2] His general conclusions seem to be that such characteristics are present for the 'general happiness' of the tribe or community. A community that is made up of thieves and murderers would hardly survive long. Somehow, virtuous behaviour is selected for because it benefits the whole tribe. Each individual needs such a benign environment in order to survive. This is possibly the only point at which Darwin is prepared to accept selection at the species level as opposed to selection acting just on individuals.

Darwin did not, of course, have the understanding of genes that we have. A modern Darwinist, such as Matt Ridley in his book *The Origins of Virtue*,[3] must assume that the basis for co-operative and self-sacrificial behaviour in humans is built on selfishness – selfishness at the gene level. We are, according to the likes of Ridley, programmed to be nice to each other.

Let me once again point out the implications of such thinking. Real love towards one's neighbour is something which most of us admire and aspire to. It is also taught by many religions. The Darwinist, however, must say that, apart from maybe the occasional freak (maladapted) person, such love does not have anything more to it than instinct. There can be no genuine free will involved, it is therefore merely an appearance and not truly genuine. This is despite the fact that the person who loves feels real heartfelt love towards someone. Such feelings are programmed, say the evolutionist, by that person's genes – for entirely selfish reasons.

Of course, genes, being inanimate structures, cannot be selfish in the sense of having any conscious motives. What the evolutionist is trying to say though is even more banal – just that inanimate molecules are determining all those heartfelt feelings of love in order that they (the genes) might propagate. This, of course, applies supremely (to the evolutionist) in the case of a mother's love for her child who carries such a high proportion of her genes. All such love, in the Darwinian world, is bogus and pre-programmed.

Can we really believe what is being said here? I am the father of four children – whom I dearly love. My love for them is not real, says the Darwinist. I am merely serving the despot inside me; the genome that screams to be reproduced at all costs, even the cost of my own life (if losing my life means preserving a sufficient number of my children). My genes are making me think that I love my children. I, the vehicle of these genes, am duped into feeling love for them.

I wonder what Darwin himself would think about this line of argument? I feel that he, like me, might rebel. Darwin suffered greatly when his favourite daughter, Annie, died young of tuber-culosis. He loved her desperately and grieved for her all the rest of his life. If such love as his was merely programmed in his genes – then why grieve? Why are we not programmed to callously for-get the dead loved one? That would surely be more adaptive for survival.

Anyone who has experienced real love, directed to us or in our hearts towards others, must argue strenuously for its reality. It is not make-believe.

I worked for some years in a remote part of Kenya, where I came to know a number of Catholic priests. I witnessed the self-giving of many of these men who had gone for life to work for poor people and lived lives of poverty themselves. They were celi-bate (how maladaptive is that?) and loved by the local communi-ty. Such behaviour is very, very un-Darwinian. How did these men come to be celibate and help people thousands of miles from their own homes who were not related in any way? I leave this question with you, the reader. It is not answered by evolutionary theory. The world is certainly full of selfishness but when we do find altruism popping up it actually rarely has anything to do with preservation of kin. Thankfully, despite war, torture and much selfishness in our world, there are still many examples of the people who live mainly for the benefit of others.

I think of that icon of twentieth-century goodness (of course, disputed by some), Mother Theresa. She left no offspring of her own. She worked for the poorest of poor people, of a different race from herself and gave her life to this end. Many continue to do the same. We admire such lives and uphold them as examples. Evolutionary principles are but dust in the face of such love.

As we pursue this matter, let us ask whether there is any actual reality to moral values. Is our so-called morality just the way our brains are hardwired to survive? If so then, why, for example, should rape be wrong? If the driving force of evolution is the gene, then that gene will have succeeded if the maximum number of possible progeny are produced from the gene carrier. Rape, whenever possible would actually be, in a Darwinian sense, highly desirable. Why, therefore should we almost universally condemn rape? I would maintain that we do so because it is wrong *per se*. It is wrong in a universal and absolute sense. Even if the Darwinist says rape is considered wrong because rape is bad for the community (and has been selected against), the implication is that what we feel is wrong about rape is merely programmed – there is no absolute morality which prohibits it.

Some modern Darwinians, you may be shocked to know, are prepared to think that such an idea of universal absolute values is nonsense. The following is an extract from 'Moral philosophy as applied science' by Michael Ruse and E. O. Wilson[4] (both highly respected giants in this field), who find it:

'easy to conceive of an alien intelligent species evolving rules its members consider highly moral but which are repugnant to human beings, such as cannibalism, incest, the love of darkness and decay, parricide and the eating of faeces. ...Ethical premises are the peculiar products of genetic history, and they can be understood solely as mechanisms that are adaptive for the species which possess them. ...No abstract moral principles exist outside the particular nature of individual species.'[4]

Here we see the depths to which the rationalist Darwinian will go. At least there is consistency. There is nowhere else for the atheist who sees nothing beyond the gene. There is a blindness here that is quite frightening.

I hope that this discussion about altruism and the examples I have given will persuade you that it is real and not mere acted out genetic programming.

The appreciation of beauty

I live in a very beautiful part of the world, in the Outer Hebrides of Scotland. As I write I see a lovely shimmering of evening light on the water of the sea loch by our home. There is a stillness and a peace broken only by the occasional call of gulls in the distance. The largest skein of Greylag geese that I have ever seen, consisting of about 100 birds has, this minute, winged its way across the view from my study. It is beautiful.

Why do humans experience great pleasure in such circumstances? Why do we appreciate art? Why do I get joy when I listen to a Bach toccata? What is the point? Is there any evolutionary reason for appreciating beauty? Is there some objective standard which says something is beautiful whether sensed by humans or not?

Let me remind you that the human brain, according to evolutionary theory has its attributes honed purely by the need for survival. There is no need for anything else. So why would the appreciation of beauty be selected for in the struggle to exist? I will maintain in this section that there is no Darwinian answer to this.

I will draw much here from the already mentioned book by Anthony O'Hear.

Even the most ancient tools of the first cavemen often have intricate carvings etched on them. These decorative designs have no functional purpose and must have been put there for aesthetic reasons only. In other words there was no clear biological advantage in having such decorations. We marvel at the cave paintings of our forebears 30,000 years ago - still vivid and beautiful. It seems that man has always appreciated beauty and created art to reflect this appreciation.

E. O.Wilson, the evolutionist, believes that aesthetic contrivances, 'play upon the circuitry of the brain's limbic system in a way that ultimately promotes survival and reproduction'. He feels that aesthetic interest is there because of the usefulness of curiosity and the search for connections and similarities.

Darwin, in *The Descent of Man*, argued that animals have a sense of beauty to aid reproduction. He noted male displays of colour, most clearly seen in the tail of a peacock. He felt that the attraction of the peahen to the tail was due to an appreciation of its beauty. There is, however, great doubt whether the peahen chooses a tail for aesthetic reasons. What we see as beautiful in the peacock's tail is more likely to be a pale stimulus to the peahen who is going on instinct alone. We shall never know for sure.

It is much more difficult to see how Beethoven's ninth symphony is customised to aid reproduction. (It seems more likely that it would take one's mind off matters of reproduction.)

Alfred Russel Wallace, who proposed evolution at the same time as Darwin, disagreed with Darwin about human nature and preferred to emphasise the differences between animals and humans. He claimed that with humans an aesthetic response is part of our 'spiritual nature'.

The philosophers Kant and Hume both said that human appreciation of beauty does not mean that there is anything real that is beautiful in a universal objective sense. O'Hear disagrees –

even if such beauty is not measurable on some scale. Some Darwinists even say that artists perform to get attraction from others, especially from women. But this totally fails to explain why the rest of us value art or performances.

The consequence of seeing beauty as an objective reality (and not merely a perception of our brains) is that the world does have some pure, transcendant something which is beyond us but real. Indeed, as Kant observed, if a flower is *really* beautiful objectively then, in a sense, nature must be organised in our interest, designed for our taste. Kant did not believe in such objective beauty (nor, apparently, in any absolute morality from without our minds).

Going back to the peahen admiring the tail; the peahen cannot really decide if the tail is beautiful or not. The peahen is programmed by her genes to recognise certain characteristics of the tail and is either switched on or off by this. On the other hand, humans can decide and debate whether something is beautiful or not.

It may not be provable but I believe that flowers are *really* beautiful and therefore their beauty has an objective reality. If not, then the rose has no more *actual* beauty than a piece of coal or a rusty nail. Nevertheless, I have every reason to believe that a rose is more beautiful. The consequences of this are, as stated above, immense. It means that there is a transcendance in the universe that is beyond us and does not depend on the firing of my brain synapses or on my genes.

If, on the other hand, there is no objective reality to beauty then the Darwinist must try hard, as E. O. Wilson has done, to see the gain to us from being so deluded. Basically, the Darwinist sees it all as a genetic con-trick designed to help us survive.

I have tried hard to see how listening to Beethoven (or the Rolling Stones) could have been programmed into my genes for survival on the African savannah (as must have been the case if Wilson is correct). I have failed to see the connection. What possible gain could I have over my fellow primates in the serious struggle for life on the plains of Africa if I have a capacity for loving Beethoven, even if the appreciation of Beethoven is partly due to particular cultural experiences in my life?

Is it not significant that the earth, when viewed from space, is like a magnificent jewel? It stands out among the planets as being stunningly beautiful. This seems to me to confirm the fact that there is an objective reality to such beauty. How could the genes of our early ancestors have been programmed to appreciate the earth from space? Likewise the pictures of the universe that we see from the Hubble space telescope show us an astoundingly beautiful cosmos, unseen by our ancestors but beautiful nevertheless. I have one such picture on the wall of my study – of the Horsehead nebula in the constellation of Orion. Its hauntingly beautiful appearance is an inspiration. My appreciation of it has nothing to do with survival of the fittest. If we are only duped or programmed to experience beauty, then such genetic programming must have arisen on the African plains many thousands of years ago. Such genetic hardwiring would have applied to the known world then – not to pictures of the earth from space (or the Horsehead nebula in Orion). All our experience of beauty confirms that what we percieve as beautiful actually is beautiful in its own right.

The philosopher Peter S. Williams has written a very useful paper on this subject and he finds that the case for objective beauty is compelling – which leads him also to see an intelligence behind the universe. I will quote from him:

'To recap, I propose the following definition of beauty: A fact is objectively beautiful if there is some ordinate degree of aesthetic pleasure that attaches to it. An 'aesthetic pleasure' is a 'disinterested' pleasure in a fact as an end rather than a means, and is the ground of characteristic behaviour on the part of persons experiencing it...On this view, beauty is not constituted by the existence of any finite mental state or states; it is not a 'sentimental gilding of reality'. The beauty of a thing does not depend in any way upon the perceiver.'

He also helpfully shows us the way in which the Genesis account of creation brings out the closeness of goodness and beauty:

'In ancient Hebrew there seems to be a linguistically enshrined recognition that goodness and beauty have close truck one with another. For example, in the book of Genesis: 'God saw all that he had made, and it was very good'. (Genesis 1:31). The word translated here as 'good' can also mean beautiful ... This affirmation resonates with the suggestion that beauty is connected to goodness and the realisation that the cosmos is overwhelmingly beautiful.'[5]

I believe that despite the horrors and pain of our world, there is a transcendance which is seen in its beauty, a transcendance that is quite beyond Darwinian theory and which speaks to us, if we would but listen, of a divine purpose.

The theory of evolution cannot explain altruism, the appreciation of beauty amongst humans nor the objective reality of beauty. All of these point to a creator and a purpose to life beyond mere survival.

1. Anthony O'Hear (1997). *Beyond Evolution. Human nature and the limits of evolutionary explanation.* Oxford:Clarendon Press.
2. Charles Darwin (1871). *The Descent of Man and Selection in Relation to sex.* Princeton: Princeton University Press.
3. Matt Ridley (1996b). *The Origins of Virtue.* Penguin.
4. M.Ruse and E.O. Wilson (1986). 'Moral philosophy as applied science', *Philosophy* 61, pp173–92.
5. Peter S.Williams (2002). *Intelligent Design, Aesthetics and Design Arguments.* Available at http://www.iscid.org/boards/ubb-get_topic&-f-10&-t-000014.html

Chapter Eleven

Darwinian Myths

Darwin's finches

When Darwin explored the remote Galapagos Islands off the coast of South America, he came across some finches which varied from island to island. In fact there are 14 different species of finch that show considerable variation in their beak size and shape, colour and body size. Yet they seemed unmistakably related and have the same song pattern and display. It is customary now to assume that Darwin used the example of these finches to derive his theory. It is stated in many textbooks that he noticed the small variations in beak from island to island and that he realised they had come from a single ancestor finch – most likely from the mainland. Over a long time it is assumed that the original finch species had diverged into the 14 separate ones – because of geographical isolation on the islands and because of different habitats and food sources. For example, large beaks would have been selected for where the food source was mainly hard seeds.

Initially Darwin did not think too much of the finches.[1] He apparently collected nine of the species and identified only six of them as finches. He also got the labels mixed up so that it was hard to know which came from which island. It was, in fact, the ornithologist John Gould who sorted out the specimens that Darwin and others provided. Darwin does not mention the finches in *The Origin of Species*. He does comment on them in his book *The Voyage of the Beagle* and there assumes that the finches may have come from a single ancestor from the mainland 600 miles away.[2] There is little doubt that, in retrospect, he saw the finches as strong evidence in favour of evolution. There are other animals which, likewise, have their own distinct species on each island, such as the Galapagos tortoises. Darwin wrote in his aforementioned journal,

'Why, on these small points of land, which within a late geological period must have been covered by the ocean, which are formed of basaltic lava, and therefore differ in geological character from the American continent, and which are placed under a peculiar climate, – why were their aboriginal inhabitants, associated, I may add, in different proportions both in kind and in number from those on the continent, and therefore acting on each other in a different manner – why were they created on American types of organisation? It is probable that the islands of the Cape de Verde group resemble, in all their physical conditions, far more closely the Galapagos Islands than these latter resemble the coast of America; yet the aboriginal inhabitants of the two groups are totally unlike; those of the Cape de Verde Islands bearing the impress of Africa, as the inhabitants of the Galapagos Archipelago are stamped with that of America.'[3]

Biologists since have found it very hard not to conclude that many of the species on the Galapagos Islands have come originally from the American continent – and that subsequently they have diversified on each island to form separate related species. Such speciation has been noted in many other places and I do not doubt it myself. This means that, contrary to the belief of many creationists, species are not in fact immutable. The big question, however, is whether such relatively 'micro' changes can account for all the variety of life that we have. I will come to this later.

The finches only achieved real status as a Darwinian icon in the 1940s – mainly because of the work of ornithologist David Lack. Later, Peter and Rosemary Grant studied the finches in great detail in the 1970s. They particularly looked at the small island called Daphne Major. Here lives the *medium ground finch.* They were able to measure and band every individual and to continue observations of climate, food and matings for some years thereafter. In 1977 there was a severe drought and the quantity of seeds was greatly reduced. The number of birds declined to 15 per cent of the original. They carefully measured the beaks of these birds. The remaining birds were slightly bigger and had an average beak width very slightly greater than the original stock.

These findings were, and still are, hailed as evidence of observed evolution in these finches. However, in 1982-83 there were very heavy El Nino rains and the food supply increased greatly – as did the population of finches. It was found that the average beak size returned to its original value. Such reversion in characteristics seems common and it is usual to see oscillating changes rather than any continuous directed change.

It was also found that in fact at least 50 percent of the species of finch could interbreed and produce successful hybrids. Strictly speaking, they were not, therefore, separate species at all. In fact some would put the real number of species at about six.

What does all this tell us? It is still a very real and impressive hypothesis that the finches did originate from a single species from the mainland many years ago. There are similar finches on the mainland of South America 600 miles to the east. We still do not, however, have any basis for extrapolating such a truth into a unified theory of evolution. This is because (as described in Chapter 7), small micro changes in characteristics, such as beak size, are due to different ratios of *existing* genes. These changes are not due to new information in the genome. New structures and forms as seen in macro evolution cannot be explained by the processes that made Darwin's finches diversify into 14 species. What is constantly cited as perhaps the strongest plank in Darwin's theory is, in fact, no explanation at all for the real macro changes that we see occurring throughout the history of the earth.

The peppered moth

Probably most biology textbooks (and certainly the one I had at school) have a section on the peppered moth (*Biston betularia*) – sometimes referred to as Kettlewell's moth or the melanic moth. Changes seen in this moth are considered one of the most important observed cases of evolution.

In the 1950s, British biologist Bernard Kettlewell performed some famous experiments which seemed to show, quite dramatically, that these moths had 'evolved' into a darker form, in order to blend in with the polluted, dark-coloured trees of cities.

The peppered moth in its typical form comes in various shades of grey with small dark flecks - hence the term 'peppered'. In 1811 it was noticed that there were some dark coloured ones, termed 'melanic'. By the turn of the century more than 90 percent of these moths around Manchester were of the melanic form. In 1896 biologist J. W. Tutt suggested that the dark-coloured moths may be due to the need to have camouflage. He noticed that the typical moth was well camouflaged against the lichen-covered trees of unpolluted countryside, whereas the melanic form seemed better camouflaged on the trees of industrial areas where the lichen was dead and the trees dark with pollution.

Kettlewell was determined to investigate this fascinating phenomenon. First, he released some of the moths into an aviary to see what happened to them. He noted that birds do eat them. Second, he released moths in polluted woodland near Birmingham. Many moths settled on tree trunks and he noted that the melanic form was much less conspicuous on these polluted trees. Third, he marked the wings of a large number of moths (both typical and melanic) and released them (during the daytime). He then collected moths from the same area at nighttime. He found that 27.5 per cent of the melanics were recovered whereas only 13 per cent of the typicals were recovered. This seemed to go well with the idea that melanic forms are less conspicuous and therefore potentially selected naturally to survive. Two years later he did the same experiment in rural Dorset where the trees were unpolluted. This time he demonstrated the opposite: the typical moths were more likely to survive.

All of this was hailed as 'Darwin's missing evidence'. We need to remember that Darwin had no evidence from observations of wild organisms to support evolution; he had relied on his observations of domestic breeding.

In the 1960s, pollution declined and it was noted that the proportion of melanic moths also declined. This was seen as further confirmation of Darwin's theory. Famous biologist Sewall Wright called this, '*the clearest case in which a conspicuous evolutionary process has actually been observed*'.[4] We should pause at this point and wonder at how a change in proportion of two forms of the same moth could be paraded as the best evidence for the process of evolution, a process that is supposed to have formed the brain, over time, from primeval soup in some warm pond.

Since Kettlewell's experiments, a number of inconsistencies have come to light. Firstly, the melanic form of moth did not replace the typicals in all industrial cities – in Manchester, for example. Secondly, in some unpolluted rural areas the melanic form predominated (and reached 80per cent in East Anglia). Thirdly, south of latitude 52 degrees north (a line just north of the Bristol Channel), the melanic form increased well after there

was pollution control. Fourthly, the proportion of melanics started to decrease in areas such as Liverpool before there was any
change in pollution. Clearly, the simple camouflage explanation
is not so simple.

It was also noted that the decline in the melanic form after pollution control was not actually associated with any noticeable
change in the lichen cover of trees. It was then found that, in fact,
these moths do not normally rest on tree trunks. They rest during
daytime on the undersides of the branches high up in the trees.
Biologists such as Kauri Miieka and Cyril Clarke have confirmed
this fact. It should be remembered that Kettlewell released his
moths during the daytime and the moths, which do not normally
fly during the day, tended to rest on the trunks of trees. Even if
placed on a tree trunk the moths would often stay there because
they are torpid and inactive in the day. Investigators became so
fixated with the idea of the moths resting on tree trunks that they
even pinned or stuck dead moths on the trunks of trees for their
studies of bird feeding. Many of the textbook photos of moths on
trees were in fact staged by sticking dead moths to tree trunks.

Where does this leave us? It shows us two things I believe.
Firstly, scientists are too easily convinced by a simple explanation
which gives credence to their theory – so much so that in the case
of Kettlewell's moths they were fixated on a false idea of moth
behaviour and an unconvincing rationale for the changes in proportions of the two forms. Secondly, however, and this must be
admitted, there were dramatic alterations in the populations of
typical and melanic forms of moth which roughly coincided with
pollution changes. Some sort of selection was occurring. Does
this amount to evolution in the sense that I understood in school?

The answer is a definite no. Even if there is natural selection
going on here, it is merely altering the proportions of already
existing stock. There is nothing new here – no new genetic information has been demonstrated. What seems to have happened is
no different from the sort of changes you would expect when a

farmer chooses certain colours to breed in his sheep. Obviously one can alter the proportions of black sheep if you preferentially choose them to survive and mate. We would not call this evolution. The alteration of proportions of typical and melanic moths is not evolution either because there is no new genetic information involved.

The Wilberforce/Huxley debate

When I saw a dramatised series on Darwin's life on TV as a boy, I was very struck indeed by the debate which occurred in 1860, at a meeting of the British Association for the Advancement of Science. The opponents were Bishop 'soapy Sam' Wilberforce and Darwin's advocate and 'bulldog', T. H. Huxley.

In the version that I saw and which is generally believed, the Bishop was portrayed as an incompetent ignoramus who arrogantly insulted Huxley by a referral to his ancestry to apes. Huxley, cool as a cucumber, was seen to trounce the Bishop with superior argument and intellect. The overall impression one got was that the Church, represented by the Bishop, was fighting a losing battle and that its arguments were futile and ignorant.

No one seems quite clear what precisely occurred.[5] Some facts are as follows: Richard Owen, the foremost anatomist of his day, had stated his claim at a meeting in Oxford of the British Association on Thursday, 28th June 1860, that anatomically (particularly concerning the brain) man was very different from any animal. He thus tried to refute any idea that there was a smooth transition from ape to man. He was, curiously, right on track with modern experts (see end Chapter 6). Huxley, at this meeting, flatly contradicted Owen and promised to put his arguments in print (which he did later). It was Owen who then primed Bishop Wilberforce to speak the following Saturday, two days later, at another meeting of the British Association in Oxford[6]. It is this Saturday meeting that is famous.

Firstly, it is worth stating that Wilberforce was not an ignorant cleric. It is hardly likely that Owen would have put anyone without credentials and ability in Huxley's line of fire. Wilberforce was, in fact, the vice-president of the British Association for the Advancement of Science. He was conversant with current science, especially geology and ornithology. He also had a first-class degree in mathematics. There is a prevailing idea that he had not even read Darwin's *Origin of Species* (see, for example, R.Tato, (ed.) (1967). *Science in the Nineteenth Century*, which actually states he had not read *The Origin of Species*). However, not only had he read it but he had already prepared a review of it which Darwin later read and admitted was a skilful criticism. In a letter to J. D. Hooker in July 1860, Darwin writes of the Bishop's review:

"I have just read the 'Quarterly'. It is uncommonly clever; it picks out with skill all the most conjectural parts, and brings forward well all the difficulties. It quizzes me quite splendidly by quoting the 'Anti-Jacobin' versus my Grandfather. You are not alluded to, nor, strange to say Huxley ... the concluding pages will make Lyell shake in his shoes ..."[7]

The Saturday meeting was very highly charged, The lecture room that had been prepared proved far too small for the audience and the meeting was held in the library of the museum, which was crammed with up to 1,000 people. An eye-witness is quoted in *The Life and Letters of Charles Darwin*, edited by his son Francis in 1888. This really was an important meeting with many of the big beasts of the scientific world present. These included Henslow, Hooker, Owen and Lyell. The Bishop spoke first, for about half an hour. He apparently used much of the material from his review article and he clearly felt that he was doing well. He seems to have been rather over-charmed by his own rhetoric and at the end he unwisely turned to Huxley and may have said, 'Was it through your grandfather or your grandmother that you claim descent from monkeys?' Huxley apparently got up and retorted, to the delight of the Darwinians present, 'Would I rather have a miserable ape for a grandfather, or a man highly endowed by nature and possessed of great means and influence, and yet who employs these faculties and that influence for the mere purpose of introducing ridicule into a grave scientific discussion? I unhesitatingly affirm my preference for the ape.' After this came a long pro-Darwinian discourse by Hooker. Wilberforce did not get up to speak again.

There is little doubt that the bishop did employ ridicule and did not help his cause in doing so. Ridicule, sadly, is used by both sides in the debate about our origins. The present-day leading popular 'bulldog' of Darwin, Richard Dawkins, uses thinly disguised, sneering ridicule in many of his writings about those who believe in a creator.

Surely though, what matters is not what rhetoric was used or how gifted in oratory Wilberforce or Huxley were. What matters is substance, and clearly Darwin did think that the substance of Wilberforce's arguments, as presented in the later review, was very impressive. This has been clouded by a populist version depicting an ignorant cleric at the mercy of science, who was totally defeated by Huxley. Please note that Huxley did not actually answer any of the points Wilberforce made in the debate but merely traded insults. To say that he therefore represents the triumph of reason in this debate is pure and absolute myth.

The debate is used in the fashion of legend, to portray a defeat of Church and a triumph of science. Nothing could be further from the actual historical truth. Many of the leading scientists of the day never agreed with Darwin and many of the leading Christians of the day did. Of the scientists who opposed Darwin's theory we can list many leading ones who were never won over by the arguments of *The Origin of Species*. These include Richard Owen, the chief anatomist of his day, and Louis Agassiz, the foremost naturalist of nineteenth-century America. Adam Sedgwick, the geologist, Lord Kelvin, the physicist and the embryologists

Von Kolliker and Von Baer all opposed the theory in one way or another. These are but a few of the many opponents and there is every reason to think that Darwin delayed the publication of *The Origin of Species* more from fear of fellow scientists than from the Church.

It is instructive to look at the very different situations prevailing at the 50th and the 100th anniversaries of the publication of *The Origin of Species*. In 1909, at the celebrations for the 50th anniversary, we notice that there was no official strong line in favour of natural selection. At the official celebration symposium in Cambridge there was a very ambiguous tone. A few selectionists stated their position – most notably Joseph Hooker and Alfred Russel Wallace who were the surviving members of Darwin's inner circle. However, there was a complete mixture of views expressed by the leading naturalists of the day and these included De Vriess, Bateson and Henry Fairfield Osborn – all of whom denied the power of Darwinian natural selection. However, at the 1959 100th celebration we see the hardened synthesis of neo-Darwinism completely dominant. I have discussed this in the chapter on the weakness of natural selection. Suffice it to say that since the 1930s there has been a hard orthodoxy which still prevails – quite different from the situation at the beginning of the twentieth century, when minds were allowed freedom to question the dogma.

Darwin's theory did not command enormous authority in the 50 years after *The Origin of Species* was published and we should be very aware of the power of myth when reconstructing the history of its acceptance.

These are but three of the commonly believed myths surrounding Darwinism. This book has shown up many others – perhaps the most blatant one being the idea that the first life appeared almost inevitably. It is not that leading scientists do not realise these mistakes. Often they do. It is just that they rarely feel free to openly say so.

1. Jonathan Wells (2000). *Icons of Evolution. Science or Myth?* Regnery Publishing.
2. Charles Darwin (1997 edition). *The Voyage of the Beagle.* Wordsworth Editions.

3. *Ibid.*,

4. Sewall Wright, *Evolution and the Genetics of Pollution*. Vol.4. Chicago: The University of Chicago Press 1978

5. See discussion in Philip Sampson (2000). *Six modern Myths*. Inter Varsity Press.

6. See also F. Darwin (ed.) (1888). *The Life and Letters of Charles Darwin*. John Murray.

7. Darwin, *The Life and Letters*, PP 324-325

Chapter Twelve

The Puzzle of Homology

One of the central planks in Darwin's theory is that of homology. It has been noted from ancient times that different groups of organisms have structural similarities which can be divided into two main types.

There is, firstly, the type of similarity known as an analogy. This is when, for functional reasons, unrelated organisms have similar structures. For example, both butterflies and birds have wings but no one suggests that this is because they have a common ancestor with wings. They both have wings as adaptations because they are useful for flying.

Secondly, there are homologies. This is when a group of organisms has a feature that seems to be the same pattern or design in all the other organisms within the same taxonomic group. The classic example is the way in which all mammals have similar bones in their limbs. We see the same pattern of bones in the wing of a bat, the flipper of a porpoise, the leg of a horse and a human. This pentadactyl (five-fingered) pattern is homology.

Darwin believed that homologies are very strong evidence for relationship based on common descent. I will quote him at length here to show his reasoning:

'We have seen that the members of the same class, independently of their habits of life, resemble each other in the general plan of their organisation. This resemblance is often expressed by the term 'unity of type'; or by saying that

173

several parts and organs in the different species of the class are homologous. The whole subject is included under the general term of Morphology. This is one of the most interesting departments of natural history, and may almost be said to be its very soul. What can be more curious than that the hand of a man, formed for grasping, that of a mole for digging, the leg of the horse, the paddle of the porpoise, and the wing of the bat should all be constructed on the same pattern, and should include similar bones, in the same relative positions? How curious it is, to give a subordinate though striking instance, that the hind feet of the kangaroo, which are so well fitted for bounding over the open plains, – those of the climbing, leaf-eating koala, equally well fitted for the grasping of the branches of trees, – those of the ground-dwelling, insect or root-eating, bandicoots, – and those of some other Australian marsupials, – should all be constructed on the same extraordinary type, namely with the bones of the second and third digits extremely slender and enveloped within the same skin, so that they appear like a single toe furnished with two claws. Notwithstanding this similarity of pattern, it is obvious that the hind feet of these several animals are used for as widely different purposes as it is possible to conceive. The case is rendered all the more striking by the American oppossums, which follow nearly the same habits of life as some of their Australian relatives, having feet constructed on the ordinary plan. Professor Flower, from whom these statements are taken, remarks in conclusion: 'We may call this conformity of type, without getting much nearer to an explanation of the phenomenon'; and he then adds 'but is it not powerfully suggestive of true relationship, of inheritance from a common ancestor?'[1]

Here Darwin puts it very well. Indeed, this bit of evidence for common ancestry for all mammals was an important factor in convincing me of the truth of Darwinism when I first studied biology. Homology is absolutely key in the edifice of evolutionary theory. What can we say about it?

The great anatomist of Darwin's day, Richard Owen, who was against Darwinism, argued that homologies were not due to common descent but more based on some sort of *archetypal* plan that the creator used in organisms. His view carries little weight today.

However, there are two main areas of study that cast doubt on a straightforward argument based on Darwinian evolution: the genetics of homologies and the embryology of homologies.

The genetic basis of homologies.

It has been demonstrated that in many cases the genetic codes for specific homologies are not, in fact, homologous themselves. We should definitely expect, if Darwin is correct, that the genes for a homology such as the limb structure of vertebrates, should themselves be homologous. The whole Darwinian argument depends on common descent and therefore common genes. If the genes for homologies are quite different then it is hard to see how they are linked by any common ancestor.

There is now strong evidence that often homologies are not in fact based on common inherited genes.[2] Back in 1971, Sir Gavin de Beer, embryologist and once director of the British Museum of Natural History, wrote[3]:

'Because homology implies community of descent from...a common ancestor it might be thought that genetics would provide the key to the problem of homology. This is where the worst shock of all is encountered...[because] characters controlled by identical genes are not necessarily homologous ... [and] homologous structures need not be controlled by identical genes.'

He concluded: 'the inheritance of homologous genes from a common ancestor ... cannot be ascribed to identity of genes'.[3]

Since he wrote this many more studies have confirmed his conclusions. Wray and Abouheif[4] write about studies, mainly done in insects, which show that regulatory genes which are homologous may be dedicated to non-homologous morphology – in other words the same genes are seen to be expressed in totally different ways, even in the same organism. For instance, the *notch* gene in the fruit fly *Drosophila melanogaster* can include the production of structures that are clearly not homologous, such as wings and bristles. This phenomenon, known as *pleiotropy*, is a common finding.

Conversely (and more relevant to this discussion), a growing number of cases demonstrate that the inverse situation often occurs – where genes that are not homologous encode a homologous morphological feature. Examples given include the genes used for sex differentiation in insects and the expression of segments in the bodies of insects. I have already mentioned *hox* genes in Chapter 7 and the puzzling way in which the same genes code for the limbs of arthropods and vertebrates.

A review of the evolution of animal limbs in the journal *Nature* describes the puzzling nature of genes used in the formation of limbs[5]. I quote from this:

'It is clear from the fossil record that chordates and arthropods diverged at least by the Cambrian. The appendages of these two groups are not homologous because phylogenetically intermediate taxa (particularly basal chordates) do not possess comparable structures. The most surprising discovery of recent molecular studies, however is that much of the genetic machinery that pattern the appendages of arthropods, vertebrates and other phyla is similar.'

There is, therefore, confusion as to what the genetic basis is for homologies. Clearly some homologies seem to share similar genes (upholding Darwinism) but many do not. Also, in the case of limbs, as already mentioned, the same sort of genes seem to control the limb development of widely separated taxa such as arthropods and vertebrates. It is a puzzle as to how such widely separated and phylogenetically distinct phyla could have the same genes for such totally different limb designs.

The embryology of homologies

One should expect, if the Darwinian position is correct, that we should see strong similarities in the way different organisms, with the same homology, develop that homology in the embryo. Once again there is confusion and puzzlement.

Denton shows that even in the earliest stages of embryo development (from egg cell to blastula stage) there are marked differences between amphibians, reptiles and mammals – all vertebrates. There are also many differences in development of homologies at later stages – for instance: the alimentary canal is formed from the *roof* of the embryonic gut cavity in sharks, from the *floor* in the lamprey, from the *roof and floor* in frogs and from *the lower layer of the blastoderm* in birds and reptiles.

The homologous forelimbs of vertebrates develop from completely different trunk segments in different groups. Another interesting case is that of the two membranes surrounding the embryos of reptiles, birds and mammals. These membranes (the amniotic and the allantoic) are considered strictly homologous but in mammals they are formed by a completely different route from birds and reptiles.

Once again, to quote Gavin de Beer:

'It does not seem to matter where in the egg or the embryo

the living substance out of which homologous organs are formed comes from. Therefore, correspondence between homologous structures cannot be pressed back to similarity of position of the cells of the embryo or the parts of the egg out of which these structures are ultimately differentiated.'[6]

I could list many more examples of cases where homologous structures are derived from completely different embryological routes – many studies have shown this in insect embryology; for instance in the totally different ways in which the alimentary canal is derived in the embryos of different species. Therefore we can see that in genetics and embryology, the case for a Darwinian explanation for homologies is very shaky indeed. It is a great puzzle.

Saltations and homologies

By now you will have realised that one conclusion that is being reached is that life has developed by a series of saltations; macro jumps in the formation of new forms and structures. Such saltations, as Darwin knew and as neo-Darwinists also know, are not compatible with evolutionary theory as we know it. Such saltations require a purposeful directive force and they require design. No random events could be able to produce saltations.

If we accept that life arose through a series of saltations then we should ask ourselves where homologies might fit in. Despite the reservations that I have shown above in the study of genetics and embryology, homologous structures still look as if they have come about through modifications from a common ancestor.

Saltations, however, do not out rule continuity over millions of years; on the contrary saltations imply that leaps of development have occurred using previous organisms as templates for the new forms. I mentioned this idea of templates in the discussion on fish/tetrapod relationships in Chapter 5. We do see in the fossil record some definite progression from simple to complex, from fish to tetrapod, from ape to hominid. Saltations can account for what we see in the fossil record as well as the presence of homologies. Homologies indeed would be expected to occur in any scenario involving saltations.

It is true that saltations are discontinuous, in that the sudden appearance of new forms is not gradual, but they also have a definite continuity in the sense that there is a history of new forms developing from older ones. Homologies would be expected in this model.

What we see, therefore, are structural homologies which imply some form of common descent (whether Darwinian or saltational) but we do not see in the genes or in the embryo, any good case for a straightforward mechanism for the homologies. Indeed, the lack of such clear genetic or embryological mechanisms points once again to some other totally different and indeed mysterious way in which life has come about.

1. Charles Darwin 1859. *The Origin of Species.* John Murray. Chapter 13

2. Wells, Jonathan (2000). *Icons of Evolution. Science or Myth?* Regnery Publishing.

3. Gavin De Beer (1971). *Homology: An Unsolved Problem.* London: Oxford University Press,

4. Gregory A.Wray and Ehab Abouheif (1998). 'When is homology not homology?', *Current Opinion in Genetics and Development* 8,pp 675–80.

5 Shubin et al (1997). 'Fossils, genes and the evolution of animal limbs', *Nature* 388 (6643), pp639–648.

6. De Beer, *Homology.*

Chapter Thirteen

Convergence

Mention has already been made in this book of the fact of convergence in nature. Convergence is when we see the same (or almost the same) characteristics appearing in organisms that are distant and unrelated.

Simon Conway Morris, Professor of Evolutionary Palaeobiology at the University of Cambridge, has written a fascinating book which is concerned very much with the enigma of convergence.[1] He has already been mentioned in Chapter 3 as one of the pioneering palaeontologists who looked afresh at the Burgess Shale fossils. His book bristles with examples of convergence and to set the scene I will mention just a few.

The eye has appeared many times in different unrelated groups of animals throughout the history of the earth. It has, in fact, basically two main forms: the compound eye, familiar to us in arthropods, and the camera eye, the type that you and I have. What is amazing is the fact that nature has thrown up these forms of eye many times in unrelated taxa. The camera eye has appeared at least seven times in the animal world. Evolutionists know that in these seven different cases there is no way that they could be related and have come from a common, camera-eyed ancestor. The closest convergence that we know for our own type of vertebrate eye is that of the octopus (a cephalopod). Other animals with camera eyes include certain annelid worms, cubozoans (a form of jellyfish) and three separate forms of snail (see figure 13.1).

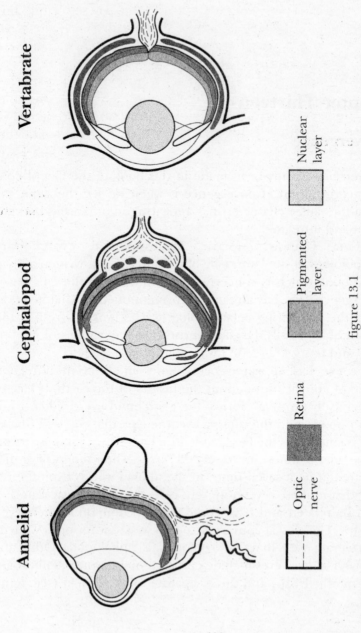

figure 13.1

Convergence of camera eye, including the classic comparison between the octopus cephalopod and human (vertebrate), as well as the alciopid polychaete (annelid). (with permission from Simon Conway Morris (2003). Life's Solution. Cambridge University Press)

The octopus eye has differences to our own. These include the position and embryological origin of the retina; and the lens, which in the octopus is formed differently and is inflexible. As you can see, however, it is remarkably like our own eye.

The phenomenon of electrical generation by fish has appeared separately at least six times and in each case involves the modification of muscle cells. The most striking convergence is between the gymnotid fish of South American waters and the mormyrid fish of the lakes and rivers of Africa. They both have systems of electrical generation and reception. By sending different types of pulse they communicate with other fish but they also create a picture of their surrounding world by sensing the structures and animals by the way these affect the electric fields. Different species have different electrical signals by which they recognise each other. I will quote Conway Morris here:

'In this electrical Babel separation of signals is, therefore, essential. What happens, however, if two fish of the same species produce their respective signals simultaneously? Potentially the result would be problematic, for the two signals would lead to destructive interference, and the whole point of signalling would be lost. To circumvent this problem the fish have evolved a mechanism that is referred to as the jamming avoidance response (JAR) in which the fish changes its signal frequency. Sensible enough, but two things are remarkable about this avoidance of jamming. First, the fish are astonishingly sensitive to potential interference. They will respond to signal modulations with an amplitude difference of 0.1% and a timing disparity of 400 nanoseconds or less, each fish shifting its signal pattern in a few microseconds. Second, and more remarkably, the algorithm used by the gymnotids and mormyrids to shift the signal has, of course, evolved independently but is identical. Nor do the convergences end there, because computationally similar neural algorithms also occur in the owl, where acuity of sensory perception is acoustic rather than electric.'[2]

As in other convergences these two forms of fish are not identical in the way they use electricity. For example, with one exception, mormyrids produce the electrical signal as discrete pulses, whereas the gymnotids produce an effectively continuous signal as a wave form.

Other convergences (and the list seems exhaustive) include the production of silk by different animals, sabre-toothed marsupial and placental cats, marsupial and placental moles, echo-location in bats and dolphins and warm-bloodedness in birds, mammals and certain fish … and many more, including hosts of convergences in the plant world.

There are various reactions to the fact of convergence. The Darwinist will say that convergence simply shows us the ability of organisms to adapt to similar environments and niches through natural selection acting on random variations. The fact that some convergences are astonishing merely makes folk amazed but does not deter the Darwinist from his or her world-view. The Darwinist emphasises the differences in the convergences, such as the different position of the retina in the octopus, and will maintain that this shows that a designer was not involved. I deal with this argument elsewhere but it is worth going over it again: to see a creator as bound to produce identical designs is simply a particular and very limited view of a creator. My own view of the creator is that, like any artist or craftsman, he will produce variety within the similar designs that are produced. A creator that is somehow limited to producing carbon copies all the time is a very limited creator.

Simon Conway Morris does seem to be coming from a Christian theistic background. He maintains (and this is really the main thrust of his book) that there is something almost 'eerie' about the way life produces convergences – that there are so many, almost infinite, other possible positions in 'morphospace' that organisms could navigate to and yet they persistently come up with the same characteristics. To him there are therefore strong hints of purpose behind it all, some deep structure or teleology within nature. His book, while upholding Darwinism, gives room for the religious biologist to breathe a little. He strongly refutes SJ. Gould's concepts of contingency[3]. Gould maintained that if you re-run the tape of life again from Cambrian times you would see a totally different set of organisms and certainly not man. Man, to Gould, is a fluke. Man, to Conway Morris, is virtually an inevitability. Conway Morris imagines meeting his first group of aliens and finding them remarkably similar to ourselves.

Conway Morris has done us a great service in cataloguing the extraordinary variety of convergences in animals and plants. He has thrown down the gauntlet to the ultra-orthodox Darwinists, the fundamentalist reductionists that we all know of. He shows that pure mechanical and contingent randomness could not produce the convergent panoply of life. Something else is going on.

I differ from Conway Morris in the way I interpret the evidence. He is wedded to the scientific answer – normal scientifically measurable processes have been responsible for evolution, despite the eeriness he mentions. I feel that he is indeed too attached to his science. As a theist (and I believe he is) he nevertheless must hold to some concept of God being active throughout. How is such activity from God manifested? If it is not simply put in place at the big bang and left to unfold (the deist position) then he must be doing his work within evolution itself. This implies and must expect the miraculous. If we limit God to only working within the laws of science then we are ruling out the miraculous. This is a hard pill to swallow – even for the Christian theist. Miracles are not part of biology as we learn and study it – they are somehow very threatening and alien. Yet, as Christian theists we should expect miracles – and if the evidence for them is there (and I hope this book persuades you that it is), then let us bite the bullet and acknowledge it. Fear of being called a flat-earther should not influence us a whit.

Convergences are remarkable, are not explicable in a Darwinian framework and point to purpose and design in nature.

1 Simon Conway Morris (2003). *Life's Solution. Inevitable Humans in a Lonely Universe.* Cambridge University Press.
2 . Ibid. P.186
3. As in Stephen Jay Gould (1990). *Wonderful Life. The Burgess Shale and the Nature of History* New York: Vintage.

Chapter Fourteen

Darwin and His Time

No scientist is uninfluenced by the prevailing beliefs and philosophies of the time he or she lives in. Darwin, though a systematic observer, was clearly influenced by many ideas which were current when he lived. It is therefore a valid and useful exercise to examine some of these when making a critique of Darwin's theory.

I believe that it is no coincidence that one of the most influential and talked about authors about evolution during Darwin's early years was his own grandfather Erasmus Darwin. As a teenager, Darwin read *Zoonomia*, his grandfather's best-known work written between 1794 and 1796. In this work, Erasmus Darwin postulated something very like the so-called Lamarckian inheritance of acquired characteristics and a form of evolutionary process similar to that of Lamarck. In Charles Darwin's own brief autobiography he speaks about his time as a medical student in Edinburgh and states:

'I had previously read the 'Zoonomia' of my grandfather, in which similar views are maintained, but without producing any effect on me. Nevertheless it is probable that the hearing rather early in life such views maintained and praised may have favoured my upholding them under a different form in my 'Origin of Species'. At this time I admired greatly the 'Zoonomia' but on reading it a second time after an interval of ten or fifteen years, I was much disappointed;

the proportion of speculation being so large to the facts given.'[1]

Here we have evidence on his own admission that *Zoonomia* probably had an influence in him writing *The Origin of Species.* Darwin himself later recognised how unscientific the speculations of his grandfather were. Nevertheless, it seems that in his most impressionable years he revered this book and, at least subconsciously, had absorbed much of its thrust.

During his period in Edinburgh he spent much time with Robert Grant, a zoologist who was an expert on sponges. He became closer to Darwin and did more to influence him than anyone else in this period.[2]

Grant was 16 years older than Darwin and a doctor who had given up medical practice to study marine life. He was a freethinker who saw nothing but physical and chemical forces involved with life. To him there was no spiritual power behind the natural world. Grant was outspoken in his praise for Lamarckian evolution and seems to have been a man of very fervently held convictions. This was Darwin's walking companion during these years. Grant felt that the very 'simplicity' of the primitive sea animals enabled us to understand more easily how more complex creatures such as ourselves could have evolved from them. He seems to have been very anti-clerical and rejected the Church's claims that the fossil record showed a series of divine creations. He was a very well-travelled man and a good raconteur – someone who kept the young questioning and impressionable Darwin very much absorbed.

It would be wrong, however, to think that Darwin took all that Grant believed 'hook, line and sinker' because we know that when he later went to Cambridge to study to be a clergyman, he firmly believed in every word of the Bible and learnt Paley's *Evidences for Christianity* by heart. The *Evidences* was the classic argument for design used by many at the time to uphold creation.

One does get the feeling, however, that Charles Darwin's understanding of Christian doctrine and his belief in it were somewhat superficial and derived from those around him rather than some deep-felt spiritual revelation of his own. It seemed to crumble without much protest when he developed his evolutionary ideas later in life.

There is little doubt that both his grandfather and Grant hugely influenced Darwin when he was young and rather without direction, thus paving the way for his later convictions about the origins of life.

Having read many books about Darwin and his own autobiography, I get a picture of a sincere young man, an avid collector of beetles and fascinated by nature, who nevertheless was highly impressionable, who had done rather poorly at school and was about to give up on the idea of being a doctor. His father was disappointed in him and worried about his obsession with nothing but shooting throughout the holidays. He courted the illustrious and learned men of the day, in Edinburgh and later in Cambridge – almost, one feels, as a compensation for his own lack of success. Even in his autobiography, written when he was 67, he seems to spend much of it recounting his meetings with the most famous scientists of the day – not exactly name-dropping but not far from it. Like all of us he was very influenced by authoritative figures around him and Grant was certainly one of these.

It is reasonable to say that Darwin's ideas on evolution did not come out of any vacuum of pure objectivity (no one's ideas do, of course) but were highly influenced by people and ideas that he had met. He himself confessed in his autobiography, with his characteristic modesty, that he was a poor critic of ideas when they first came to him. There is almost a naivety in his acceptance of the thoughts of others. Here is a fascinating excerpt from that autobiography:

'I have no great quickness of apprehension or wit which is

so remarkable in some clever men, for instance Huxley. I am therefore a poor critic: a paper or book, when first read, generally excites my admiration, and it is only after considerable reflection that I perceive the weak points. My power to follow a long and purely abstract train of thought is very limited; and therefore I could never have succeeded with metaphysics or mathematics. My memory is extensive, yet hazy: it suffices to make me cautious by vaguely telling me that I have observed or read something opposed to the conclusion which I am drawing, or on the other hand in favour of it; and after a time I can generally recollect where to search for my authority. So poor in one sense is my memory, that I have never been able to remember for more than a few days a single date or a line of poetry'[3]

Darwin then goes on to say that, because of this, some of his critics have said he was a good observer but had no power of reasoning. He refutes this by referring to *The Origin of Species* which he famously describes as '*one long argument from the beginning to the end*'.[4] There is no doubt that *The Origin of Species* is a carefully argued book, honed over many years of preparation. But this does not mean that it was a purely objective piece of reasoning. Probably no scientific theory is. However, it seems that in the case of Darwin there was much more in his background to influence his ideas than with most scientists. I would maintain that he already had strong inclinations towards evolutionary theories long before he really put his ideas together after his voyage on the *Beagle* and that these were strongly influenced by the essentially atheist philosophies of Erasmus Darwin and Robert Grant. This would explain why he latched onto facts to support the theory, such as geographical distribution of species. We now know (or should) that such small changes in groups of organisms (such as the Galapagos finches) are not due to any new information in the genome but to reshuffling of the existing alleles – yet such small variations were a key foundation of *The Origin of Species*. Likewise, we now know that that other foundation for his theory, artificial selection in domestic breeds, is not due to new genetic information and cannot account for the macro changes of the fossil record. He, understandably, held onto these illusory foundations for his theory because he had already become convinced of the idea of evolution. Even Huxley balked at extrapolating the findings of pigeon fanciers to claims about the evolution of, for example, the eye. Darwin, however, once his course was set, could not break away from the rigid rules of gradual, imperceptible change that he had set himself and preferred to apply the findings of domestic breeders to his theory rather than see its limitations and go down what he perceived to be some saltationist alley.

Darwin was, in one sense, a genuinely humble man but, paradoxically, there is strong evidence that he was also extremely ambitious and determined to make a name for himself in the world of science. It is worth looking at this aspect of his character because it must have influenced the way in which he saw his theory and the acclaim it would bring. He did not know for sure that it would bring acclaim and indeed it suffered much criticism – but in the end he was at the top of the scientific establishment, the very thing he had dreamt of.

That he was very keen to make a name for himself comes through in his autobiography. When at studying at Cambridge he continued his childhood pastime of collecting beetles. One specimen was very rare and his name was associated with its discovery. He describes his feelings in the autobiography: 'No poet ever felt more delighted at seeing his first poem published than I did at seeing, in Stephens' 'Illustrations of British Insects," the magic words, "captured by C. Darwin, Esq."[5]

Later in the autobiography he admits to being ambitious – to gain notoriety amongst top scientists, although not the public. When on the *Beagle* he heard that his former Cambridge geology teacher, Sedgwick, had been to see his father and said, 'that I should take a place among the leading scientific men.'[6] He later heard that this was because of some scientific letters he had written home, which had been published. He was also very excited by the fact that his collection of fossil bones sent home from the *Beagle* had attracted considerable attention amongst palaeontologists. He writes:

> 'After reading this letter, I clambered over the mountains of Ascension with a bounding step, and made the volcanic rocks resound under my geological hammer. All this shows how ambitious I was; but I think that I can say with truth that in after years, though I cared in the highest degree for the approbation of such men as Lyell and Hooker, who were my friends, I did not care much about the general public.'[7]

True, such ambition is common to many, but I feel that his need to be approved of by his father and the establishment was very great. We must remember that his mother died when he was only eight and he could remember nothing about her. His highly successful father dominated his early years and until the success of the researches he had made on the *Beagle*, he had not shone at all. When a great theory about nature came to him, however gradually, it must have been intoxicating to think that he had solved the whole meaning of life's grandeur. It is conjecture to imagine how this may have coloured his objectivity but it is at least worth our scrutiny.

The principle influence on Darwin when on the *Beagle* voyage was undoubtedly the work of Charles Lyell, the great geologist. Darwin brought with him the first volume of Lyell's great *Principles of Geology* (just published). He read this avidly and was delighted to receive the second volume later in the voyage. Lyell's 'uniformitarian' approach to the history of the earth undoubtedly provided Darwin with a framework in which he could imagine the gradual progression of evolution. Lyell painted a picture of a very ancient earth that had stayed much the same (the opposite of catastrophism). The sediments had been laid down steadily over millions of years. Such a steady picture of the earth over long periods gave succour to the idea of steady progression in formation of species. Lyell himself was not, at that time at least, an evolutionist. Later, he was a close friend of Darwin but remained sceptical. Geologists now know that that both catastrophism (volcanic activity, major climate changes, asteroidal impacts etc.) and uniformitarianism are both part of the whole history of the earth.

The fact that Darwin was reading Lyell's work at the time of the initial formation of his theory has great significance. Had he known that the earth had regularly suffered catastrophes of climate and various impacts – with large-scale extinctions – it is possible he might have had a less certain geological foundation for his ideas.

Every new theory in science swims in the sea of general beliefs and philosophies of the time. As such, these beliefs provide a support both for the author of a new theory and those who agree with it.[8]

Isaac Newton was, par excellence, the epitome of the great scientist and the model for how scientists should see the world in the nineteenth century. Darwin, along with much of society, was much influenced by 'Newtonianism'. Darwin, in his species notebooks, constantly followed the example of the major scientific philosophers of the time (Herschel and Whewell) in seeing all science in the light of invariable natural laws – as in Newtonian astronomy. Darwin had read with great interest a review of Comte's *Cours de philosophie positive.* This work, in true Newtonian spirit, emphasised that all phenomena are subjected to invariable natural laws. At the end of 1938 Darwin reread the major works of Whewell (*History of the Inductive Sciences*) and Herschel (*A Preliminary Discourse on the Study of Natural Philosophy*). While not evolutionists they also admired and expounded the Newtonian principles and the idea that science would somehow solve every mystery. When Darwin first read Herschel's *Discourse* at Cambridge he had been entranced. In his copy of the book Darwin had scored a passage:

'To what, then, may we not look forward ... what may we not expect from the exertions of powerful minds', building on the 'acquired knowledge of past generations?'.

The most influential philosopher in England at the time was John Stuart Mill, whose *System of Logic* was published in 1843. He had enormous influence, again strongly in favour of absolute laws of nature and universal causation. He wrote: 'it is a law that every event depends on some law'.

This philosophic environment of inviolable laws and chains of causes influenced Darwin greatly. To Darwin, and many other thinkers, the appearance of life in all its forms must, like in Newton's astronomy, be under entirely natural and rational laws that could be discovered and described. The idea of miracle therefore was anathema to this way of thinking (and still is). Darwin wrote the following words to his friend Lyell on the subject of miracles:

'If I were convinced that I required such additions to the theory of Natural Selection, I would reject it as rubbish ... I would give nothing for the theory of Natural Selection, if it requires miraculous additions at any one stage of descent.'[9]

We need to see where Darwin was coming from – once again we cannot uphold the idea that his theory was developed in some vacuum of objectivity. On the contrary, it fitted admirably with the undercurrent of scientific philosophy of the first half of the nineteenth century.

There is a misconception that religious thought was all against

Darwin's theory. However, it is important to realise the effect that German 'higher criticism' of the Bible was having, not just on the continent but also to a degree in England. There was a significant liberal branch of the Church that questioned the authority of the Bible and which doubted anything miraculous. Such liberalism was against any form of teleology in the appearance of life. It therefore suited Darwinism well and certainly helped to prepare the ground for it. Darwin was influenced by such theology and read works such as *Phases of Faith* by Francis Newman (John Henry Newman's brother). This was at the time of his daughter Annie's terminal illness. In this book Newman outlined an emotional, spiritual odyssey from orthodoxy to severe doubt. It echoed Darwin's own journey and he described the book as 'excellent' in his notebook. It is hard to know how much the subsequent tragic death of Annie affected his faith – and later the bitter blow when Amy, his son Frank's wife, died after childbirth. It would seem that these painful experiences of loved ones dying did nothing for any belief in a caring or benevolent creator.

There was, in Victorian Britain, a great tension amongst thinkers and theologians between the traditional view of God as a benevolent gentleman figure who cared for every detail of his creation and the facts of biology such as the extinction of species, the apparent waste of life and the bloody, seemingly heartless struggle for existence that lay behind the pastoral façade of nature. Darwin seems to have felt this very acutely and throughout his writings he continually refers back to this picture of a benevolent God which is incompatible, to Darwin, with reality. His theory of evolution, maintains Cornelius Hunter in his book *Darwin's God*[10], was a form of theodicy – a way of accounting for the problem of evil by distancing God from his creation.

One of Darwin's favourite books was *Paradise Lost* by John Milton. He brought his copy of this book on the voyage of the *Beagle*. The thrust of *Paradise Lost* is to solve the problem of evil with an emphasis on God letting humans choose between good and evil so that the good can be separated from the bad. Milton's God was pure but passive, distanced from the events of history. Such a view of God allowed one to see creation on its own, rather than under God's influence.

In his autobiography Darwin wrote about suffering and natural selection; underscoring his basically metaphysical basis for evolution:

> 'Suffering is quite compatible with the belief in Natural Selection, which is not perfect in its action, but tends only to render each species as successful as possible in the battle for life with other species, in wonderfully complex and changing circumstances.'[11]

That the theory of evolution is founded on metaphysical beliefs about God is important and instructive. The fact that it is founded on ideas about how God *ought* to behave should lead us to look hard at the sort of constrained view of God that was prevalent in the nineteenth century. Even now we see apologists for evolution constantly bringing God into their arguments. Richard Dawkins, and many others, pepper their arguments with concepts about how God should behave if he created us – particularly when speaking about suffering in nature. Hunter, in his book, shows that such a metaphysical foundation is not actually science – more a way of coping with one's own belief system. As he writes:

> 'From Milton to Liebniz, Hume, and others, modern intellectuals were rapidly moving away from the view that God creates and controls the world and toward the view that God must be separated from evil. In the nineteenth century these views would play an important role in Darwin's development of evolution. The common denominator between Darwin's evolution and the earlier theodicies is that God governs via secondary causes – his fixed natural

laws – and that God is justified to humankind when we view natural evil as a result of some sort of cosmic constraint outside of God. Darwin worked with this tradition, and it is no surprise that he arrived at his theory of evolution, which claims that nature's imperfections and evils arose from natural forces rather than a divine hand.'[12]

To reiterate: Darwin was no exception – all theories are coloured by some degree of subjectivity. I maintain, though, that the theory of evolution, much more than others, is founded on very strong subjective elements: his own grandfather's influence, Grant and his freethinking atheism, Darwin's own powerful ambitions, a prevailing nineteenth-century philosophy of Newtonian naturalism and, perhaps most importantly, an emerging Victorian idea of what God should or should not do.

1 F. Darwin (ed.) (1888). *The Life and Letters of Charles Darwin* John Murray, P.38

2. See Adrian Desmond and James Moore (1991). *Darwin*. Penguin.

3 Darwin, *The Life and Letters*, P.102

4. Ibid.

5. *Ibid.*, p.p. 50-51

6. *Ibid.*, p. 66.

7. *Ibid.*, p. 66.

8. See Michael Ruse (1999a). *The Darwinian Revolution*. Chicago.

9. Darwin, *The Life and Letters*.

10. Cornelius Hunter (2001). *Darwin's God*. Brazos Press.

11. Darwin, *The Life and Letters*.

12. Hunter, *Darwin's God*,

Chapter Fifteen

The Watchmaker

I have chosen to write a detailed critique of a book by the Oxford professor Richard Dawkins called *The Blind Watchmaker*.[1] This is because Dawkins remains the most vocal and prominent Darwinist of our times. I have mentioned his name a few times already. He is immensely successful in that he has published a number of best-sellers, all on the theme of evolution. He specialises in arguing against design and God. Possibly his most quoted and well-known book amongst the public is the one I am going to look at in depth.

I believe this is a fair way to end my book because if I cannot out-argue Dawkins then there is not much point in spending time writing all of this. I believe that those who are interested in this debate would want to see some sort of a full-blooded alternative to Dawkins. He does invite this because he is deliberately provocative in the way he belittles any opposing ideas. He is a powerful writer who seems to argue effectively and is completely committed to his cause – indeed I respect him for his ruthless and undeviating approach. However, all this makes him a quite a good target for some logical and scientific criticism. I intend to undo his arguments chapter by chapter, beginning with his preface.

Preface

He begins with the following words: 'This book is written in the conviction that our own existence once presented the greatest of all mysteries, but that it is a mystery no longer because it is solved. Darwin and Wallace solved it ...'[2]

I hope that anyone who has read my book so far will feel somewhat baffled that Dawkins can be so sure of the fact that the mystery is 'solved'. We have no idea how life began, we do not have fossil evidence of Darwin's gradual evolution, mutations are not the answer to the sudden appearance of new body plans, natural selection is an inert process that cannot produce the complexity of life, biochemical systems are seen to be irreducibly complex and human nature defies an evolutionary explanation. These are just a few of the problems with Darwinism that I have discussed at length here.

Dawkins, in the same preface, states that:

'More, I want to convince the reader, not just that the Darwinian world view happens to be true, but that it is the only known theory that could, in principle, solve the mystery of our existence.'[3]

We will study his arguments in detail but it is very interesting that he sees Darwinism as the *only* possible solution. He seems unable to accept that any other way could account for our existence. I believe much of this is because of a type of 'scientism' which dominates his mental landscape. Anything which seems outside of scientific explanation (such as God) must not exist. Miracles are ruled out. The supernatural is out of the question.

I would maintain that such a view is in itself a sort of faith. It is a belief in the absence of the supernatural which has no empirical evidence to support it. It is, in a way, a form of religion, sometimes called atheism.

He dismisses critics of Darwinism as ignorant:

'Darwinism, unlike "Einsteinism", seems to be regarded as

fair game for critics with any degree of ignorance.'[4]

Well, let us see how his arguments measure up to the real evidence for or against Darwinism.

Chapter 1. Explaining the very improbable.

In this chapter Dawkins explains that we, and other life, are very complex. We give the appearance of having been designed. He contrasts the complexity of life with that of physics, which he says is relatively simple – despite the difficult mathematics it uses.

He discusses man-made objects, such as aircraft, which have been designed. We are much more complex and, he states, have not been designed. He says that Darwin showed us that we are not designed objects. At this stage he does not give any reasons for this but simply states it as a fact.

He discusses the famous treatise by the eighteenth-century theologian William Paley who wrote Natural Theology – or the Evidences of the Existence and Attributes of the Deity Collected from the Appearances of Nature, published in 1802. This was, at one time, a favourite book of Darwin and Dawkins himself admires it as a careful and knowledgeable work. Paley argues cogently from the appearance of design in the multitude of systems in nature. He famously describes the possibility of finding a watch lying on a heath. On finding such an object one would automatically deduce that it had a maker because its mechanism shows compelling evidence of design. He likens the exquisite complexity of the parts of life, such as the eye, to the watch on the heath. With many examples Paley passionately argues for evidence of God as the great designer.

Dawkins then states that natural selection has been found to be the answer to life's complexity and that it is like a *blind* watchmaker. Unlike any idea of God, it is unconscious and without any purpose or foresight. Paley, he states, is 'wrong, gloriously and utterly wrong'.[5]

Dawkins gives away no reasons yet for believing this. It is just stated as a fact. He then says: 'Darwin made it possible to be an intellectually fulfilled atheist.'[6]

He then spends some time trying to explain what sort of complexity life has. He agrees with Paley that it has the appearance of design and that we must find answers that satisfy as to why this is. He gives the name *hierarchical reductionism* to his approach to understanding life's complexity. We are complex at different levels – all dependent on each other; from molecules to cells to organs to bodies.

He probably sums up this chapter with this pure Darwinian statement about how he sees the emergence of life: 'We shall explain its coming into existence as a consequence of gradual, cumulative, step-by-step transformations from simpler things, from primordial objects sufficiently simple to have come into being by chance.'[7]

He ends the chapter with an overview of the incredible complexity of the eye, complete with diagrams of the eye, the retina and one of the photo cells. He reminds us that each of the many cells contains a nucleus with a database larger, in information content, than all 30 volumes of the *Encyclopaedia Britannica* put together.

What is heartening is that Dawkins at least takes the fact of life's amazing complexity very seriously. He knows that it needs a serious answer.

He is convinced that the answer is that of Darwin. However, thus far, he has not given us a single reason for such a view. It seems that he thinks that by saying it often enough, with all the authority he can muster, he can persuade the reader before even starting to unpack his reasons. The book has 11 chapters – so there is hope that such reasons will be forthcoming.

Chapter 2. Good design

Dawkins starts with: 'Natural selection is the blind watchmaker, blind because it does not see ahead, does not plan consequences, has no purpose in view.'[8]

He then writes that in this chapter he will look at a particular example of complexity in nature that would impress any engineer. He says that he will then consider possible solutions to this problem and that he shall finally come to the conclusion that nature has adopted.

There follow 15 pages of detailed and awe-inspiring description of the way in which bats use sonar to 'see'. I admired the way in which Dawkins dealt with this. The description is indeed quite stunning. We learn, for instance, about the 200 pulse per second ultrasound output with a send/receive switch system – so that the bat is not sending sound at the same time as receiving the reflected sound. We learn also of frequency modulation – where the bat makes each pulse of sound output of varying pitch, like a wolf-whistle. This means that the reflected sounds received at any time will be of different pitch from the output, thus preventing jamming of the system. Engineers have dealt with similar problems in designing radar. It is wonderfully complex and intricate. He also describes how bats use the 'Doppler effect' to measure the velocity of insects or other prey. We get the impression that bats use sound to build up a picture of the world which may be just as detailed and even 'colourful' as our own visual world. Dawkins is well worth reading if only for his ability to describe such magnificent aspects of nature very concisely and beautifully. He concludes:

'I hope that the reader is as awe-struck as I am, and as William Paley would have been, by these bat stories. My aim has been in one respect identical to Paley's aim. I do not want the reader to underestimate the prodigious works of nature and the problems we face in explaining them.'

Then he explains that the basis for most people's objections to Darwinism is what he terms the 'Argument from Personal Incredulity'.[9] Just because we are amazed at life's complexity we should not jump to any conclusion that it is designed. This is a fair point. Pure amazement is not enough; there must be rational, analytical and logical reasons for objecting to Darwinism.

He spends a little time more or less 'writing off' a book by Hugh Montefiore who (like a modern Paley) tries to argue for a designer of life. I have not read this book and it may well be that Dawkins' criticisms are valid. He seems, however, to have picked on a particular book that has little serious scientific underpinning. He enjoys pointing out some inconsistencies and more or less ditching the entire argument because of this.

He then explains that most objectors to Darwin do not appreciate the immensity of time needed for changes to have evolved. He points out that people are prepared to believe that change occurred in the melanic moth (see detailed analysis of this in Chapter 11, 'Darwinian Myths') and he implies that if such a change (to darker moths) occurred in just 100 years then we should not be surprised if great changes occur in millions of years. Here Dawkins is just plain wrong. As I have already explained, and as most geneticists know, there is no new genetic information involved in the story of the melanic moth. The genes for the darker form were already in the moth population. For Dawkins to use this as an example of Darwin's small step- by-step change shows either a lack of knowledge about genetics or a plain refusal to see the truth.

He then, to back up his point, gives the example of dog varieties. Over a few hundreds or thousands of years we have (through artificial breeding) got the St Bernard and the Chihuahua. These, he admits (one feels rather reluctantly), remain dogs but he says they are an example of what can happen over a relatively short time. He says that if we extrapolate this to several hundred million years then we should not be surprised at the evolution of all the different types of life. Here Dawkins once again seems to show a lack of understanding of genetics. What has happened in dog varieties is akin more to micro-evolution. Micro-evolution, as I have repeatedly stated, occurs because of well-understood differences in the distribution of existing alleles (not new ones). This is exactly why dogs remain dogs no matter how hard you try to select varieties. This subject is dealt with well in Raymond and Lorna Coppinger's book *Dogs*[10]. They go into some detail about the genetics of dog varieties. Some characteristics, such as short legs, may be the result of mutations (similar to achondroplasia in humans) but the rest is pure recombination of existing wolf genes. If Dawkins could point to a case where, by artificial selection, a species has changed into an entirely different one (not a variety) with entirely different characteristics and which cannot interbreed – then he might have a point. But he does not do this because it has never happened.

He refers again to the time scale. He says that if the time from wolf to modern dog is represented by one pace then from 'Lucy' (described in Chapter 6) to humans is two miles. He again misses the point. There is nothing radically new in the dog breeds – simply reshuffling of the alleles within the dog genome. Raymond and Lorna Coppinger's book is fascinating on how alterations in growth rates of different parts of the dog skeleton provide the great variety of shapes – particularly the differences between muzzles of different breeds. The same genes are there; it is more a case of how they are used. In all this we see nothing that amounts to more than changes in size or colour; nothing new is seen on the macro level. In contrast to the dog lineage, the supposed transformation of 'Lucy' into *Homo sapiens*, however, would require immense amounts of new genetic material. We have already covered some of this ground in showing that many modern geneticists are talking about 'macro-mutations' and 'saltations' to achieve the unspeakably enormous changes needed to give modern *Homo sapiens* just the power of speech.

His closing arguments involve the need to appreciate the immensity of time available for evolution, the fact that natural selection is not random (he gets annoyed when objectors balk at the idea of random events causing complexity). He then gives the case of the cuckoo. The cuckoo lives a parasitic life that involves laying eggs in other birds' nests; also the baby cuckoo has the habit of throwing the hosts' own chicks out of the nest. Dawkins quotes CE. Raven, who maintains that these characteristics of the cuckoo are essential conditions for its parasitic way of life – and that each on its own is useless. Raven says the odds against evolution providing both characteristics at once are too large. Dawkins rightly criticises this particular argument because, in fact, it could be seen that even one of the characteristics on its own could benefit the cuckoo. Here Dawkins has chosen an easy target – an ill thought-out creationist piece of writing. He would not however find it so easy to explain the 'irreducible complexity' of, say, the feather or the bacterium's flagellum or the haemoglobin molecule. He attempts to do so with the haemoglobin molecule in the next chapter but, as we shall see, fails.

Chapter 3. Accumulating small change

This is where at last we hope to see some concrete arguments for the sort of small gradual changes over time that Darwinism requires.

Firstly, he shows us that nature does often, on its own, throw up ordered systems. His example of this is the pebbles on a beach. As you walk along a beach you see that the pebbles are not arranged randomly. There are segregated zones with small pebbles and large pebbles separate. The ordinary forces of nature involving the action of waves have done all this. He imagines a primitive tribe living near the beach who see this order and attribute it to a Great Spirit in the sky. He shows how wrong this tribe is.

Another example is that of a hole. If objects above the hole are shaken around then only ones that fit through it can go below it. Thus, there is a sorting of the objects by size. This, he explains, is another example of essentially blind physical forces creating order. Another example is that of the planets going around the sun. For any given orbital distance there is only one speed that a planet can travel at to stay in that orbit. This is not designed – it merely appears so because it is ordered. He sees such systems as 'sieves'. Sieving in nature gets things in some degree of order (the pebbles, the objects with the hole and the planets).

Dawkins is right that nature creates systems of order without anything other than unconscious physics involved. Where he is wrong is in extrapolating such order to be the cause of meaningful information. We discussed this in Chapter 2 on the origin of life. There is all the difference in the world between semantic (meaningful) information and syntactic (meaningless) information. The snowflake is ordered and beautifully so. It has no meaningful information in it however. The pebbles on the beach are also ordered but have no meaning. In contrast, the sequence of nucleotides in DNA is full of meaning and purpose – it determines the characteristics of an organism. All the examples that Dawkins has given of naturally ordered systems fail to impress because they are essentially without any other purpose or meaning and have no other reason for being what they are. Life is so different.

He then gives the example of the haemoglobin molecule and tries to explain how it could have come about in small random steps. He tackles the whole problem of how small changes in the amino acid sequence of a protein can lead to this molecule which carries oxygen in our blood. As he says, the haemoglobin molecule consists of four chains of amino acids twisted together. Each chain consists of 146 amino acids. It is helpful to quote from him as he introduces the subject:

'There are 20 different kinds of amino acids commonly found in living things. The number of possible ways of arranging 20 different kinds of thing in chains of 146 links long is an inconceivably large number, which Asimov calls the 'haemoglobin number'. It is easy to calculate but impossible to visualise the answer. The first link in the 146-long chain could be any one of the 20 possible amino acids. The second link could also be any one of the 20, so the number of possible 2-link chains is 20 X 20, or 400. The number of possible 3-link chains is 20 X 20 X 20, or 8,000. The number of possible 146-link chains is 20 times itself 146 times. This is a staggeringly large number. A million is a 1 with 6 noughts after it. A billion (1,000 million) is a 1 with 9 noughts after it. The number we seek, the 'haemoglobin number', is (near enough) a 1 with 190 noughts after it! This is the chance against happening to hit upon haemoglobin by luck. And a haemoglobin molecule has only a minute fraction of the complexity of a living body.'[11]

Dawkins then explains that simply adding amino acids on to a chain randomly is never going to achieve a haemoglobin molecule. Up to this point we are in complete agreement. He then says that it is through cumulative selection that it can be possible to construct a haemoglobin molecule over many generations of an organism. The idea is that if even a small amount of the molecule is made, then it will be selected and will be reproduced until one day it becomes a little bit more like haemoglobin and then this is selected, and so on. He then set up a computer program to test his idea. He chose a line of Shakespeare, 'Methinks it is like a weasel', and set his daughter, aged 11 months, on to a keyboard of a computer (restricted board with 26 capital letters and a spacebar only). The computer was set up so that when a letter or space was put in correctly (for the above sentence of Shakespeare) then that character was kept in the line. The sentence has 28 characters in it and so his daughter (a suitable ran-domising device) kept typing lines of 28 characters. Dawkins then checked to see how many lines it took for her to get the entire sentence right. He ran the test three times and it took 43, 64 and 41 times to achieve the sentence. From this he concludes that cumulative selection, as simulated in this computer trial, can build complex molecules over time, much sooner than could be expected if it was totally random and non-cumulative.

There is a fundamental error in this reasoning, as I will explain. An organism that supposedly starts out making a haemo-globin molecule does not have any picture or program of the final chain. In the computer model Dawkins has entered in the exact sentence at the start and the computer 'knows' when his daughter hits the right character. There is no such system in an organism. His idea is that when a chain of amino acids is ran-domly started that somehow the environment 'knows' that it is needing haemoglobin – even if there are another 145 amino acids to get in place. This is, of course, totally incorrect. There is no blue-print that nature has for a haemoglobin model (not unless you are ready to concede that there is a divine one). In the random world of Darwinism there may indeed be selection, but selection of what? A tiny fraction of a haemoglobin molecule? Such a fraction of a molecule would already need to be an effi-cient oxygen carrier (and releaser) at the very start in order to have any selective advantage.

Actually, Dawkins merely emphasises the incredible fact that there is anything like a haemoglobin molecule at all and the idea that it has arisen through random mutation is quite untenable.

Quite amazingly Dawkins then agrees that his model is invalid. He writes:

'Although the monkey/Shakespeare model is useful for explaining the distinction between single-step selection and cumulative selection, it is misleading in important ways.

One of these is that, in each generation of selective 'breed-ing', the mutant 'progeny' phrases were judged according to the criterion of resemblance to a distant ideal target, the phrase METHINKS IT IS LIKE A WEASEL. Life isn't like that. Evolution has no long term goal. There is no long-term target, no final perfection to serve as a criterion for selection, although human vanity cherishes the absurd notion that our species is the final goal of evolution. In real life, the criterion for selection is always short-term, either simple survival or, more generally, reproductive success.'[12]

Dawkins has just demolished his own model and says that life is not like that. He is quite right. His computer model was a total waste of time. Why then did he spend so much time telling us about it? It is very hard to understand how Richard Dawkins can abide having this example of a computer program in his book. Has he removed it from his latest editions – if not, why not?

He then describes a computer program which he designed to show evolution more graphically. He calls this a computer ana-logue. By using a branching tree-like pattern he introduces 'mutations' to see what happens. The program consists of simple line drawings which branch or fork with each 'generation'. If left on its own it forms a bushy tree-like figure after a number of gen-erations. He introduced nine different alternative ways (or genes) for a branch to grow (in length, angle to other branch etc.). By allowing such mutant genes to occur randomly there are soon a wide variety of possible shapes. When he saw a shape that inspired him for any reason then he singled it out to continue branching (stopping the others, which became extinct). This is how he put a form of natural selection into the process. By doing this he obtained many interesting two-dimensional shapes – some of which resembled insects or birds. Dawkins feels that somehow this represents how evolution must happen – a system of growth and generations with random changes and some form of selection to choose the more successful varieties. He calls the successful shapes 'biomorphs'. He then suggests that there are, in fact, a fixed number of potential biomorphs in this system. This is because essentially he is using a logical, rule-based program that, can only result in a finite number of possible shapes. He extrapolates this idea to say that, in the realm of real life, there are also a fixed number of potential organisms 'out there'. We see only those that have somehow been selected from the random mutations over the years. He seems quite unabashed at the sense of exhilaration he had when he first saw the various biomorph shapes appear on his computer screen.

My first reaction is that this is not really science as I know it. A computer program to simulate evolution may seem a neat idea but it is not empirical. It is not based on actual observation of the real thing and is, if you like, merely conjuring up a second-hand imitation. As such it has little to teach us and is based on theory rather than hard facts.

Secondly, the 'biomorphs' that are produced may resemble some sort of insect or animal but for Dawkins to then somehow liken such simple, two-dimensional, non-functional line drawings to the reality of life seems almost childish in its naivety. These shapes have no function or meaningful complexity. They cannot 'do' anything. The information within is meaningless – syntactic rather than the semantic stuff of real life. The fact that they resemble insects is entirely due to the way he chose to follow certain patterns that he liked. It frankly astonished me that today's leading exponent of Darwinism should use such a crude and unconvincing computer system to try to persuade us of his case.

Once again we are waiting for some real evidence to get our teeth into.

Chapter 4. Making tracks through animal space

Dawkins starts this chapter with a re-examination of the eye. He acknowledges that many people find it very hard to believe that the eye could have come about through small steps with no designer. He poses two questions:

1. Could the human eye have arisen directly from no eye at all, in a single step?
2. Could the human eye have arisen directly from something slightly different from itself, something that we may call X?[13]

The answer to the first question is, he says, a decisive no. The odds against a 'yes' answer are enormous, he says ('many billions of times greater than the number of atoms in the universe').[14]

Please note that his entire frame of reference here is that there is no deity or designer – so his answer *has* to be no. He has not argued thus far to convince any serious enquirer that there is no designer. He just assumes we are with him on this atheistic point. I find this quite telling. He writes as if it is a fact that the reader entirely shares his own belief system. So, having demolished, in his view, any idea that an eye could appear ex nihilo, he tries to answer question 2. Not suprisingly his answer is yes.

He does this by explaining that if X is something very nearly like a human eye then the human eye could plausibly have arisen from X by a tiny alteration in the genetics of X. If we accept this then we could accept that X itself could have arisen from a more primitive, fractionally different X ...and so on. He uses this line of reasoning to say that the eye could indeed have arisen by small step changes without any design. I quote:

My feeling is that, provided the difference between neigh-

bouring intermediates in our series leading to the eye is sufficiently small, the necessary mutations are almost bound to
be forthcoming. We are after all, talking about minor quantitative changes in an existing embryonic process.[15]

(In Chapter 9 I have already given the reasons why such an idea
of eye evolution is false. For the sake of clarity I will, however,
repeat some of those arguments here to refute Dawkins.)
   This is where Dawkins loses his concentration and gets it quite
wrong. He states here that the only changes required are *quantitative*. If this is what he believes then he must also believe that the
most primitive, original eye, must have had no *qualitative* difference from a modern eye. In other words, it must have had a cap-
sule, eye muscles, cornea, iris, lens, retina (including the molecular machinery to convert photons to meaningful electricity) and
a nervous system to interpret it all. These are the most obvious
qualities of a human eye. He is trying to make ourselves, and himself, believe that the mutations needed are simply a little bit more
or a little bit less of what is already there (quantitative changes).
This is, of course, nonsense. To achieve an eye there must have
been endless mutations to give entirely new, *qualitatively* different
characteristics. Let us look again at the iris, for example. It must
have arisen somehow *de novo* and it must have acquired many new
characteristics of immense complexity such as: its position in
front of the lens, its shape and ability to open and close, its nervous system connections which intricately adjust the aperture in
response to varying light and its other nervous system connections which adjust its aperture for accommodation (when we look
at something near the pupil gets smaller). These are not a little
bit of quantitative tinkering. These are very major complex *new*
characteristics.
   To therefore say that it is possible to imagine a long series of
Xs in smooth transition with almost undetectable differences
between successive ones, is preposterously wrong. There is noth-

ing smooth about the acquisition of a lens or a superior oblique muscle (one of the series that control eye movement). These are major innovations which cannot be acquired simply by changing what is already there. When I read this sort of nonsense I get angry. I am amazed that he gets away with it. Wake up everybody!

He then spends some time in explaining how even 5 per cent of an eye can be of use. This is to provide a rationale for the belief that very primitive eyes (seen in some organisms) are beneficial and so have the potential for improvement over a long time. Here, some of his reasoning is sound. Of course, even 5 per cent of an eye, - such as a mere light-sensitive spot, seen on some single-celled animals, is of use.

There are indeed, in the animal world, various visual systems which look very like intermediates between a primitive sort of eye and our own. There are eyes that have no lens, there are eyes that have no cornea – merely an opening such as a pinhole camera has. This is indisputable. However, it is completely wrong to say that this means we can overcome the logical difficulties of the various parts coming together by random mutations to achieve a human eye. He actually has no idea how, for example, a mutation could give the *Nautilus* a lens. The *Nautilus* is a relative of the squid and octopus. They, unlike the *Nautilus*, both have lenses and eyes very similar to our own – though they have come about entirely independently. Dawkins rarely expresses doubts or worry but he does so concerning the lack of lens in the *Nautilus*. I quote:

'What is worrying about the Nautilus is that the quality of its retina suggests that it would really benefit, greatly and immediately, from a lens. It is like a hi-fi system with an excellent amplifier fed by a gramophone with a blunt needle. The system is crying out for a particular simple change. In genetic hyperspace, Nautilus appears to be sitting right next door to an obvious and immediate improvement, yet it doesn't take the small step necessary. Why not? Michael

Land of Sussex University, our foremost authority on invertebrate eyes, is worried, and so am I. Is it that the necessary mutations cannot arise, given the way Nautilus embryos develop? I don't want to believe it, but I don't have a better explanation. At least Nautilus dramatises the point that a lensless eye is better than no eye at all.'[16]

He is right to worry and he displays here again some of the naivety that underpins his belief system. He actually says that the change needed is 'simple' and just a 'small step'. This is again (forgive me for the strong language) utter nonsense. To acquire a lens is not simple. A lens is not just a blob of transparent material somehow placed within the eye. Lenses are very complex – made of a unique material which is both transparent and flexible and which is surrounded by muscles which alter its shape in response to very complex nervous inputs. It is marvellously constructed to iron out chromatic aberrations. Even if we agree that an imperfect lens could be of some use, there would have to be a minimum requirement for it to be of any benefit. Such a minimum requirement involves too much for us to believe in any random process arriving at it. There actually is no way in which we can see a lens just appearing by chance one day in a mutant *Nautilus*. I throw out a challenge for anyone to explain to me how a *Nautilus* could, by a single genetic mutation, acquire a functioning and beneficial lens.

He speaks then of the so-called 'convergent evolution' that has led different animals to evolve eyes very like our own. I have discussed these in Chapter 13. The prime example is that of the octopus whose eye is virtually identical – though it cannot have obtained it through a common evolutionary path. (In fact, as I mentioned before, there are seven separate animals that have independently acquired camera-eye lenses.)

Though very different, the compound eye of the trilobite (described in detail in Chapter 3) is, in its own way, hugely complex and appears instantly in the fossil record during the Cambrian period. The fossil record does not lend support to Dawkins' idea of gradual eye evolution.

He has another go at a creationist writer who describes the problem of the bombardier beetle being able to evolve in steps the two chemicals it uses to 'explode' when it squirts them at an enemy. His argument against the creationist is simply a negative one – in nothing he writes does he give any solid reason to believe gradual evolution has actually occurred. He betrays his irritation and superior feelings when he calls such writing 'Anti-evolution propaganda' and 'pathetic'.

He spends time then discussing the case of lung evolution. Once again, he describes how a primitive organ (such as a lung) could be of benefit even if only partially formed. He tells us about the lungs of certain fish that can exist outside water and maintains that this proves that a staged process of lung evolution is feasible. I would agree that it shows us how 'primitive' lungs can be of benefit. Where he fails is in persuading us that every stage is a random event leading up to advanced mammalian lungs. I discuss the appearance of land animals and the supposed evolution of these from fish in Chapter 4.

He then 'launches' into a discussion about the evolution of wings in small incremental steps. Nothing in his arguments moves me away from my conclusions about this subject written in Chapter 5 of this book.

He lists rather randomly various other organs that could, in his opinion, simply have arisen gradually by chance – including the ear, bat echo-location and the venom of snakes. All of these examples are thrown at us with no empirical data to support his views. It seems that once he has decided that random mutations + natural selection + long periods of time = gradual evolution – then anything is fodder for his neat formula. He just has to say it happened and he seems convinced. Once again – no hard data, just theorising and story-telling.

He claims confidently that any organ or apparatus that we actually see is the product of a smooth trajectory – all we have to do is imagine the series of Xs (as described with eye evolution). Once again he applies a simplistic idea of quantitative changes in tiny steps leading to complex organs. The logic is wrong; as I have stated with the case of the eye, complex organs require many *qualitative* novelties to have appeared. Tinkering with what is already there does not produce anything – all along his imaginary trajectory new genetic information is needed – the sort of information that provides the iris of the eye.

He spends some time in describing certain flat fish such as plaice, sole and halibut. They require flatness in order to live on the sea bottom. To achieve a flat position they are like other fish except they are on their side. This poses a problem if one eye is then always facing downwards. In these fish the young are vertically flattened like other fish. Later they settle on the bottom, on their sides and the lower eye migrates onto the same side of the fish as the other one. The skull of the fish also twists around to allow for its new orientation. Dawkins sees this as good evidence of evolution. You can see his point and it is an easy one to make. It looks like the original fish ancestor was like other fish but then some mutations allowed it to lie on its side and have a twisted head plus migrating eye. He calls the whole anatomical arrangement distorted and imperfect – betraying its ancient history of step-by-step change. I take issue with him on two major points. Firstly, there is nothing imperfect or distorted about the plaice or the sole. Simply because there is a different way of orientating the skull and eyes does not spell imperfection. On the contrary, they are very well adapted to their life on the bottom of the sea. Secondly, once again he is postulating how this arrangement could have come about gradually without any empirical data. He gives no fossil evidence of intermediates and he does not explain how gradual stages could have conferred survival advantage. For instance, presumably the gradual steps included fish that were not fully on their sides and fish that had only a partially migrating eye. One should try to imagine what advantage these had. I cannot and I am unaware of any fossil evidence to back up his claims. We are left again with pure conjecture – hardly good science.

He then describes how the human eye is imperfect (in his opinion) because the photo cells in the human eye are orientated in a strange way. The nerves from the photo cells come out of the side nearest the light and travel across the front of the retina before going into the optic nerve. Dawkins maintains that this is a poor arrangement because it must compromise the amount of light that gets to the photo cells. It would be better, he feels, if the wiring was on the side away from the light. Such apparent imperfection leads him to see this as a product of evolution and not of a designer. No engineer, he feels, would dream of designing the eye in this way.

Once again I believe he is talking with remarkable ignorance here. Simply because he cannot fathom why the wiring is this way does not make it wrong or imperfect. The proof is in the pudding – in this case the amazingly good vision that we possess. There is nothing imperfect about it. There is a sense here of an amateur looking at something he cannot understand and then saying it must be imperfect because he cannot understand it.

He then spends time discussing the phenomenon of convergence. Convergence, as outlined in Chapter 13, is when entirely different groups of animals separately acquire features that are almost identical. As we have already seen there are many examples and he helpfully describes a few of them. One is the octopus eye, which is almost the same as the human eye (though interestingly the nerves of the retina travel behind the photo cells, unlike in humans). He mentions echo-location again – it appears in bats, some birds, dolphins and whales. He discusses electric fish – there being two forms of the weak electric type, entirely separate in the 'new' and 'old' world. I mention all of these in Chapter 13. There are many examples of animals which, whether extinct or not, have seemed almost identical – though from entirely different continents and therefore which cannot have shared the same history.

The classic case is the similarity of many marsupials in Australia to mammals elsewhere, the Tasmanian wolf, the marsupial anteater and the marsupial mole being good examples. There were marsupials in South America long ago and an amazing case of convergence is the marsupial sabre-toothed *Thylacosmilus* which was almost identical to the sabre-toothed tiger (*smilodon*) of the 'old world', which we know and love.

He lists others but perhaps most striking is the case of cicada insects. They have a very brief adult stage and then the juvenile nymphs live underground for many years. Some varieties stay 13 years underground and other varieties 17 years – before emerging in unison. The astonishing thing is that this timing has been reproduced in at least three entirely separate species around the world, which apparently could not have 'evolved' this habit together. They all seem to have picked 13 and 17 years as somehow crucial to survival – quite independent of each other.

He spends considerable time on the various 'trades' amongst animals – particularly concentrating on burrowers, large hunters and plains grazers. In the three main separate areas of the world ('old', 'new' and Australia) there have been animals in all these trades which strongly resemble those of the other continents. A good example is that of the extinct litopterns in South America, which were uncannily like horses.

Dawkins, like Darwin, sees all this as compelling evidence for evolution. The fact that in all these cases of convergence there is never an absolutely exact resemblance, carries weight with him. This, he maintains, shows that any idea of a designer is wrong. The minor differences between these convergent animals show, he says, that they evolved separately. It is the similarity in environments of these animals which, he says, has determined their remarkably similar needs and thus convergences.

At the risk of repeating myself from Chapter 13, I have two main points to make about convergences. The first is that it seems to me actually very likely that if there is one designer then that designer would use similar themes throughout the world. It is no argument to say that the minor differences between convergent animals betrays a Darwinian evolutionary past. In the case of human designers there is every reason to expect similarities between one design and another. For example, a designer of cars will tend to carry on using similar though different designs for the various varieties. When I see a Rolls Royce I usually can tell it is one. There are different Rolls Royces but they all have a certain 'Rolls Royceness' about them. This is because the design teams in Rolls Royce like to continue similar themes. This is the nature of designers. The individual cars have not 'evolved' – on the contrary, they are quite separate entities which are similar only because of the designers. In the same way we usually can tell a Van Gogh from a Renoir – because these artists had certain traits and preferences.

It seems therefore to me absurd to say that the appearance of a marsupial mole and an 'old world' mole requires us to ditch design as a reason. On the contrary, it speaks to me of a designer who uses similar themes and likes to repeat them (though always a bit differently) wherever he wishes. It would amaze me if the designer used absolutely identical designs in different regions – just as it would amaze me if Van Gogh had reproduced absolutely identical paintings.

Secondly, it is the very, very extraordinarily close resemblances that speak to me of common design. It is very difficult to believe that the almost identical sabre-toothed tigers of the 'old' and 'new' worlds were simply due to similar environments. They are so alike that they inspire one to wonder and ponder about what common thread binds them. The environments that they lived in were somewhat similar – but it is very hard to believe that similar environments can produce animals so alike that the facial (and dental) characteristics are so incredibly the same. They look as if they are from out of the same design shop. I believe they were. Dawkins and others want to give the environment too much control over the characteristics of the species – down to the last facial and anatomical detail. I believe they are stretching our credulity to expect us to believe that similar, separate environments will throw up such strikingly similar beasts. Remember that all the characteristics of the sabre-toothed tigers, in the Darwinian world, came about by accidental mutations with no design. I do not believe that in such a scenario we would see almost identical proportions and features. There are so many ways of hunting prey – a large hunter can have many possible anatomies and still be successful. Why, therefore should we see such amazing convergence? Design seems to me the obvious answer. I have already mentioned that other astonishing example, the octopus eye. Apart from very minor details it is the same as the human eye – yet it cannot have evolved from a common ancestor. To expect such closeness to have occurred by random mutations is unacceptable.

He finishes this chapter with descriptions of the army ants of South America which parallel the behaviour of driver ants of Africa. Once again we see remarkable convergence. The same arguments as I have used for the other cases can be applied to these ants.

This chapter once again shows the paucity of evidence that he has to buttress his arguments for Darwinism. His discussion about the tiny incremental changes needed to form the eye are illogical and fail to convince. The arguments about convergent evolution simply betray a prior belief in his creed of Darwinism. In fact, such parallel similarities around the world speak more clearly of a common designer.

Chapter 5. The power and the archives

In this chapter Dawkins explains how DNA works. He gives an overview, not dissimilar to that which I have written earlier in this book. He explains the digital form of information and the extremely accurate copying mechanisms of DNA down the generations.

He states what has become a very important part of his understanding: that 'living organisms exist for the benefit of DNA rather than the other way around'.[17] This is a theme that recurs in his writings and which underpins his philosophy and view of life. It is a view that makes us and all other organisms simply vehicles for DNA – there is no other purpose or point. This stark and very limiting understanding fuels his atheism. There really is no point in it all apart from competing DNA.

He then describes some experiments to show how mutations can cause change which is inherited and beneficial. His example is that of some studies of RNA in laboratories. The RNA was in solution (not in cells) and allowed to freely replicate itself. When a poison to RNA is introduced in small quantities then the RNA which can resist it gets to survive and is replicated in the subsequent batches of RNA mixture. Increasing the concentration of the poison causes further selection and yet more resistant RNA is found to be selected.

This experiment is supposed to make us believe that new information leading to new structures can likewise be selected for and reproduced in life. As such it fails to convince. (It is very interesting that the cases of evolution most often quoted as having been observed are to do with organisms becoming resistant to drugs or poisons.) Resistance to a toxic agent is not in the same semantic realm as forming some new structure (such as part of an eye). Resistance to an agent is to do with tinkering with the molecular structures that are already there (such as the cell wall of a bacterium in resisting an antibiotic). There is no brand new meaningful information. All that is happening is merely alteration of some aspect of the existing molecular structures so that a toxic agent is unable to attack. This is a defensive and rather inert process. There is nothing innovative about it. To therefore use such examples as arguments for evolution is completely false. Once again, Dawkins (along with many others) fails to think through the logic of his arguments. Acquiring resistance to a toxic agent is light years away from what we should expect of Darwinian evolution. I would be more impressed if the experimenter found that some RNA had formed a cell membrane around itself – or some proof-reading enzymes to prevent copying mistakes. You can be sure that the first trilobites on the floor of the Cambrian ocean did not come about by mere tinkering.

He then describes an imaginary scenario involving a mutation in the DNA of a beaver. In a rather extraordinary section he imagines how this mutation will change the coding for a protein which affects the beaver's neurological transmission in a part of its brain such that its behaviour is altered. The beaver might then hold its head higher in the water when carrying logs for its dam. This is advantageous because more mud remains on the log which can then stick to the dam. The dam is more effective and therefore the beaver's offspring survive better. This behavioural characteristic is selected naturally and the new genetic information gradually replaces the older. I find this extraordinary because it assumes a direct link between one mutation and a beaver holding its head higher in the water when involved in a specific task. There is no evidence or understanding that allows us to say such a thing is possible. We really have no idea how many factors are involved in changing such instinctive behaviour. This is glossed over. Once again we are left with a sort of hopeful, rather naive assumption that mutations can do all that we expect them to. Once again we have no empirical evidence whatsoever.

One might have expected that this chapter about the very basis for inheritance, DNA, might have shed more evidence for Darwinism. It seems to have done the opposite.

Chapter 6. Origins and miracles

Dawkins begins this chapter with a discussion about what he thinks a miracle is. He maintains that what we call miracles are nothing but very improbable events. His main thesis is that an event such as the origin of life on earth is very very improbable but not impossible. When we have both the enormous time available plus the fact that the universe consists of vast numbers of galaxies and stars – then it can be shown that a very improbable event such as the manufacture of DNA could conceivably happen by chance, even if it only happened once ever in the universe.

He admits that there is a problem in imagining how the first replicating molecular machinery could have come about. I quote:

> 'The theory of the blind watchmaker is extremely powerful given that we are allowed to assume replication and hence cumulative selection. But if replication needs complex machinery, since the only way we know for complex machinery ultimately to come into existence is cumulative selection, we have a problem.'[18]

Later in the chapter he tries to give some ideas about how the first replicating molecules might have started but at this point we need to challenge his definition of a miracle. He is entitled to his particular belief system about lack of supernatural events but he does not have the right to force anyone else to accept it. My *Oxford English Dictionary* defines a miracle as: 'an extraordinary and welcome event that is not explicable by natural or scientific laws, attributed to a divine agency'. That is what a miracle is. The 'scientism' of Dawkins rejects this and he attempts to persuade us that a miracle is simply very, very improbable but always under normal scientific laws. His world-view cannot take in anything that is not within those scientific laws. This is what underlies his entire book. He does not actually give any evidence anywhere that he is right – it is his belief system.

He then pours scorn on those who suggest that a creator might have initiated life on earth. He says this is a 'transparently feeble argument, indeed it is obviously self-defeating'.[19] He goes on to say that if a creator did create the first life then the creator must be very complex and organised, particularly if 'we suppose him additionally capable of such advanced functions as listening to prayers and forgiving sins'. [20]He says that invoking a supernatural designer is to explain precisely nothing, for it leaves unexplained the origin of the designer. He says: 'You have say something like "God was always there", and if you allow yourself that kind of lazy way out, you might as well just say "DNA was always there" or "Life was always there", and be done with it.'[21]

We take note of his language about those who believe in a creator – feeble, self-defeating, lazy. This passage fairly bristles with his righteous indignation. It actually betrays his problem most clearly. His problem is that his own belief system is being challenged by those who believe in God. His lack of understanding of theology is stupendous. It has always been an accepted doctrine of the Judaeo-Christian faith (and of many others) that God is eternal and not made. In other words we do not need Richard Dawkins telling us that this is what we have always believed. That God has been always God and never created is central and made quite clear in the Bible. That Dawkins cannot accept this is his own difficulty but it does not give him the right to rubbish those who do accept it. His reasons for rubbishing such orthodox beliefs is quite unscientific. He has no evidence at all that God could not have always existed. He does not even try to give any. It is just that within his very limited frame of reference he cannot tolerate the idea.

Even if we do not accept the biblical concept of an eternal uncreated God, there is still a noble philosophical history of logical reasoning for the existence of a creator. Since Aristotle, the idea of a 'first mover', who must have been always there simply to account for matter as it is, is a valid and powerful metaphysical argument for God. Dawkins, however, does not even attempt a refutation of such arguments.

He then spends some pages going into the idea that an extremely improbable event such as the spontaneous generation of life (what he calls the spontaneous generation probability or SGP) can be imagined if we give enough time and enough available planets in the vast universe. His point here is valid to a degree, in that we do need to take on board the enormous time scales and numbers of opportunities. He is right that our human brains are not built to take in such scale. His arguments, however, are inconsistent. He says something like this (I am paraphrasing his argument):

> The spontaneous generation of life is an incredibly improbable event but the opportunities for it to occur are vast in number given the huge universe and the long aeons of time and therefore we can accept that it happened by a random mechanism. The formation of an eye in one evolutionary step is an incredibly improbable event but this is not consistent with Darwinism so, despite the long aeons of time and vast numbers of planets in the universe we cannot accept it as happening by chance in one step.

You see he wants it both ways. He is right that the universe and its time scale give scope for improbable events. He cannot, however, simply pluck the spontaneous generation of life out of a list of other improbable events and say that it could have happened by chance while other improbable events could not have. I will give another example. It is conceivable in scientific terms that by pure random chance all the nitrogen in the earth's atmosphere and all the oxygen (every last molecule) could separate totally into two pure gases which are not mixed in any way. Not only this, it is scientifically possible that this state of complete separateness could last for one hour. Does Dawkins think that this has actually happened somewhere in the universe? Obviously not. What I am driving at is that he seems not to realise that the spontaneous formation of life is (as described in Chapter 2 of this book) as far as we can see, just as improbable. To therefore out rule a miracle (a real one) is not up to him. All his arguments about probabilities fall very short of any evidence that life started spontaneously by pure chance.

He then spends a few pages discussing one scenario relating to how the first replicators could have begun. He uses the work of Cairns-Smith to describe how various clays (and the crystals of these) could have a sort of replicator effect on the environment by improving their own chances of being deposited in water systems. I will not go into detail here but I did find this an undeveloped and naive assumption – that clay crystals could form some sort of meaningful scaffolding for the start of replicating DNA. It is all conjecture and once again fails to account for any meaningful information in such a replicator system. It has never been reproduced in a laboratory nor observed in nature and there is no hard evidence. As such it is not real science.

The entire chapter could be summed up as saying: the start of life is very improbable but the universe is very big and there has been a long time for it to happen by chance. I am not clear why he took so long to say this. It is valid as an idea but is hardly much evidence and does little to convince us.

Chapter 7. Constructive Evolution.

He starts this chapter by accepting the fact that many people see natural selection as a purely negative force, 'capable of weeding out freaks and failures, but not capable of building up complexity, beauty and efficiency of design'.[22] In fact, I am one of these people and have tried to explain my reasons in Chapter 9 on natural selection.

He maintains that there are two main reasons for seeing natural selection plus mutations in a much more creative and constructive light. The first is the concept of 'coadapted genotypes' and the second comes under the name of 'arms races'.

The idea of 'coadapted genotypes' is simply that any gene has to work in tandem with others. A gene cannot affect an organ in the body unless that organ already exists. That organ depends for its existence on a host of genes all working in a co-ordinated fashion. A gene cannot work in isolation but must 'collaborate' with other genes. This all makes good common sense. The success or failure of any mutated gene would entirely depend on whether it can fit into the existing order of genes to confer some advantage. Once there is a particular arrangement of genes; for example, for some chemical pathway, then any new gene must fit in with that existing system. Similarly if an animal already has a particular way of living then any evolution will tend to keep to that same path. An example he gives is that of carnivores and herbivores. Any new advantageous gene in a lion will not be in the direction of eating grass – because the lion is already far down the path of carnivorous adaptation.

We can see the logic of all this. But ... we need to look again at the original objection that natural selection is a purely negative and uncreative force. Why does this idea of gene co-operation help the case of natural selection? This remains entirely unclear and Dawkins does not help us to see anything in the system of coadapted genotypes that rescues natural selection from such a criticism. On the contrary, I would maintain that the idea of genes having to co-operate all the time within a very complex multi-gene genotype is entirely restrictive for natural selection. It means that any mutation that occurs must stay within the game plan of the existing genes. This actually rules out novelty and would inhibit the formation of completely new structures. We see again the problem that evolution has in explaining the eye. How did the new genes for a completely new structure (such as the iris) get a look in at all? Perhaps even more amazing is, how did the new genetic material for the neo-cortex of man, involved in speech, get incorporated into the existing genetic collaborative system?

He then digresses to discuss the genetic material known as introns. As I explained in Chapter 7, introns are lengths of DNA occupying a large percentage of chromosomes which (until very recently) had no known function. They are not genes and are, in fact, spliced out of the template when RNA is reading the code. Dawkins suggests that they are the remnants of previous ancient DNA that has been deleted (as in a computer deletion). He says that these introns may hold 'fossilised' genes that have been superseded. I also discuss the latest research on introns in Chapter 7. This research is (not surprisingly) showing that introns are very important indeed and have essential functions which we have hardly begun to understand. Dawkins wrote his book before such findings but betrays again his non-teleological position by assuming that introns are just 'junk'.

He also mentions the very much accepted idea of Margulis that mitochondria and chloroplasts were originally bacteria introduced into primitive eukaryotic cells about 2 billion years ago. The idea is that bacteria entered these cells and then were somehow incorporated into the cell as proper constituents. I have many doubts about this interesting idea but that would take another entire chapter.

Having digressed, and thus far not persuaded me that natural selection is either creative or constructive, he then tackles the issue of genetic 'arms races'.

In an arms race, say between two world powers, there is constant competition to keep abreast or ahead of the enemy. A bomb on one side is made bigger than that of the other side – and so on. A missile tracking system is developed in response to enemy missiles and the missile makers make devices to jam the tracking equipment etc., etc. Often a balance of technology is reached whereby increasing sophistication on each side does not lead to one being more dominant.

In the same way, it is argued, the cheetah and the gazelle are in an 'arms race'. The cheetah needs to go faster to catch gazelle and the gazelle needs to get away quicker. The idea is that such a race encourages (through natural selection) greater sophistication. This is why Dawkins uses the idea to say that natural selection is not uncreative and negative – rather it is able to be creative because of the competition amongst enemies such gazelle and cheetahs.

This is a superficially neat argument. There is little doubt that a faster cheetah that catches more gazelle is more likely to have more offspring and be 'selected' to pass on its genes. Its slower brother may not survive or may not sire so many offspring. This makes sense. Such a process will tend to weed out the weaker and encourage the stronger to continue. But, and this is an enormous but, we still have to account for the source of the variation that is providing the better-equipped cheetahs. In micro-evolutionary terms this is easy; there is already some variation in the gene pool of the species and so those genes that are more successful will be passed on. Small differences within the species will be selected to continue – just as a farmer breeds from the more productive animals. But what about macro evolution? How do we account for the big changes that we know have happened? Variation in the gene pool is not enough for that. Once again we come to the question of whether mutations can provide the macro-changes that have led from, say, an early shrew-like mammal to a blue whale. Natural selection, even in the arms race scenario, is merely a bystander waiting for the right variations to arise. It has no power in itself to influence what those variations are. Darwinian evolution requires that the elephant's trunk and the bat's echo-location arose entirely by chance – only after the variation arises does natural selection decide whether it succeeds. Dawkins' arms race does absolutely nothing to solve this problem for Darwinists.

Natural selection is still in fact, a negative and uncreative force even in the arms race. The weaker are killed off – that is all it can do.

Dawkins, of course, spends over 20 pages of his book going into all this but essentially this is all it amounts to. His two rescuers for the beleaguered natural selection ('coadapted genotypes' and 'arms races') fail entirely to save its reputation.

Chapter 8. Explosions and spirals

This chapter is basically about sexual selection. This is the well-known phenomenon of when the members of one sex have a choice over which mate to have. The classic example, repeated in many other species, is that of the peahen who selects the male peacock of her choice depending on the size and magnificence of his tail.

The interactions of female preference for a large tail and the utilitarian need for a smaller tail are very intricate. There is a utilitarian need for small tails because large tails are cumbersome and prevent escape from predators.

Dawkins summarises the quite demanding complexity of these interactions and goes quite a bit into ideas of positive feedback and runaway evolution. These go some of the way into explaining the size of the peacock's tail. All this is interesting and neat. I will not try to detail it here because he does a very good job of it. Also it does not really help to argue the case for or against Darwinism – as I shall explain.

His main subject of discussion is not the peacock but the African long-tailed widow bird. It has been shown experimentally in the field that females do indeed prefer males with longer tails and so there is clearly a selection pressure on males to have longer tails.

Next comes the complicated bit (you may need to read this a

231

few times): because longer-tailed males will have had mothers who preferred longer tails (because its father would have had a long tail) then the longer tailed-males will also carry (unexpressed) female genes for longer-tail preference. Therefore the female choosing the longer tail is also choosing genes of her own type of preference as well as the genes for a long tail. Hence the positive feedback and increasingly long tails – which, however, are pulled in the opposite direction by utilitarian selective pressures to be shorter.

There is one main reason why none of this impresses me as far as explaining the peacock's tail is concerned. The reason is that none of this elegant theorising about the tug of selective pressures can explain the beauty of the tail. We can just imagine that the length of the tail (as in the long-tailed widow bird) can be altered by the process of female selection. We can imagine also a preference randomly occurring in females for a colourful tail. What we cannot begin to imagine is why females should have randomly developed a particular liking to the stunning iridescent patterned tail of the peacock. Surely a nice colour would have done the trick; not the work of art that we actually see. What we see is overkill, riotous overkill. It seems downright unnecessary in a purely Darwinian world. The peacock's tail is justifiably considered one of the marvels of nature. Its aesthetic appeal is beyond question. It is magnificent. My question is ... why? I know that the Darwinist will give an answer along the lines of the female choosing (over millennia) more and more colourful tails with all the positive feedback they can muster. I do not buy it. Sadly, we cannot enter the mind of a peahen. What is she looking for? Could her desire for colour have somehow, over millions of years, fostered the tail we see? I believe the answer is no. Remember what I have already stated a few times: in Darwinian evolution, any novelty in nature, including the tail of the peacock, had to arise (however gradually) by entirely random mutations. All selection does is weed out the unsuccessful. This is what happens when a peahen does not choose a male. That is all. Selection is a sieve – and as such is passive. All novelty and all creativity has to appear independently and randomly prior to being selected. That is how we need to see the peacock's tail. Did it come about by such a random and directionless process or was it designed?

Perhaps you as a reader will not be too impressed by these arguments. I have not proved that the peacock's tail is designed; but then neither has the Darwinist proved his case. The ball is really in the Darwinists' court.

Chapter 9. Puncturing punctuationism

I have already made brief mention of a school of thought that sees the evolution of species (from one to another) as going in jumps. The palaeontologists Stephen Gould and Niles Eldredge coined the term 'punctuated equilibrium' in 1973[23] when they first wrote about this idea. They saw, as others also see, that when we look at the rock strata we notice that for long periods, usually many millions of years, there is no change of note in any given species. This is known as 'stasis'. Then, suddenly, geologically speaking, one sees a change to a new species occurring – followed by prolonged stasis, and so on. This appears to be a pattern which is actually recognised now by most palaeontologists. Gould and Eldredge also extrapolate from this the idea that natural selection works on the species just as much as on the individual. Somehow the species is selected and changed as one unit – not just the sum of individuals being selected.

It must be emphasised that this idea of punctuated equilibrium is not the same as saltationism (which I have also discussed). Gould (who died recently) and Eldredge are still Darwinists and believe in evolution. They would not subscribe to any idea of massive macro leaps, such as the sudden appearance of an eye from nothing. Nevertheless, they do refine Darwinian theory and challenge some of its sacred tenets – particularly they challenge the ideas that all evolution is entirely gradual and that selection only acts on the individual.

Creationists have sometimes leapt on this idea and made it seem as if punctuated equilibrium is tantamount to separate acts of creation by God. Gould and Eldredge have absolutely denied this. However, this is one reason why Dawkins wants to write about the subject. He does so to try to rescue Darwinism from any taint of creationist thinking.

In this chapter he defends the notion that Darwin and neo-Darwinists have that evolution is actually entirely gradual and in the next chapter he writes against the concept of species selection being of any importance. As such he crosses swords with Gould and Eldredge. Though reasonably polite (at first), he ticks them off for indulging in some sort of mythology. Gould and Eldredge, based in the United States, are giants in the world of palaeontology and evolutionary theory – so Dawkins' attacks upon them have not exactly gone down well across the Atlantic. The resultant 'Darwin Wars' have been the subject of much rancour and books have been written about the fall-out of this dispute.

It is important to read Gould on this subject. In particular I would recommend his tome The Structure of Evolutionary Theory[24] and 'The Interrelationship of Speciation and Punctuated Equilibrium'[25]. You will find that Gould was one of the most erudite expositors of evolutionary theory around and his knowledge of palaeontology was prodigious. It is hard not to side with him in this debate, simply because he seemed to know so much more than Dawkins.

Dawkins uses an analogy of the Children of Israel going through the desert from Egypt to the promised land. If looked at in detail, their movements from camp to camp were erratic and unpredictable. At times they stood still but eventually they reached their goal. On average they travelled 24 yards per day. If looked at purely as an average speed then you might think that they crawled 24 yards each day before setting up camp again. The reality was, as I have just said, much more erratic and 'punctuated'.

Dawkins likens the way in which the Children of Israel travelled to the way evolution is seen in the fossil record. What we see is often long periods of stasis followed by sudden appearance of new forms. The record of speciation is jerky.

He becomes genuinely confused in this chapter and seems to say on the one hand that evolution is gradual (not punctuated) and yet on the other hand if looked at carefully is jerky. He swings from discussing the extreme stasis of the coelacanth fish to the apparent (but he says illusory) suddenness of the Cambrian fauna. He takes a swipe at creationists using the Cambrian fauna for their arguments but he gives no reasonable explanation for the Cambrian explosion – the only explanation he mentions being that perhaps the precursors of the Cambrian animals were soft bodied and therefore not preserved. Chapter 2 on the Cambrian explosion shows that this is quite untenable as an argument. He mentions the extraordinary tripling of brain volume in 3 million years from *Australopithecus* to *Homo sapiens* but gives no reason for it.

In a somewhat paternalistic way he chides Gould and Elredge for courting controversy about punctuation. He says that they are really gradualists themselves if you scratch hard enough.

His main reason for explaining the apparent jumpiness of the fossil record is that of migrations. What he means is that when a small part of one species gets isolated and geographically separated from the parent stock, then speciation can occur in the small separated group – I have discussed this fully in Chapter 7. Later, the new species may migrate back into the territory of the parent stock and eliminate the original species. The idea is that the fossil record of the original species will show only stasis followed by a sudden change to the new species (even though evolution has been gradually occurring all along somewhere else).

This seems reasonable at first glance but on closer scrutiny the argument is very forced indeed. It presupposes that all the cases of punctuation that are seen (and they seem to be the main picture, in micro-evolution at least, throughout the fossil record) are due to migration of a separated stock back into the area of their ancestors. This could hardly be the case for every single time we see punctuation. Gould denies that this migration is the reason for punctuation because he points out that in many cases the parent stock does not die out but carries on alongside the suddenly appearing new species. Gould sees instead a genuine branching in the evolution of species – not the migrations and extinctions that Dawkins advocates.

Dawkins then tries to maintain that, in fact, there is no reason why the ancestral stock in itself cannot evolve without migrations into it from a previously separated stock. He maintains that there is no inbuilt inertia within a large species group and tries to get rid of any idea that only small separated groups can evolve. This goes against his previous argument that punctuation is due to migrations back into the ancestral area. He is trying here to have it both ways – to provide a plausible explanation for punctuation (migration) while still refusing to imagine that there is anything about the original parent species that inhibits evolution occurring without migration. This seems to be because he does not like any idea that the species in itself is selected rather than individuals within it.

If you are confused about all this then you are in good company. Certainly Dawkins confuses the issues constantly and there is little cohesion in his arguments.

He also spends time in this rambling chapter to attack saltationism. Saltationism as you will remember, is the idea that major changes have occurred suddenly – in what seems like miraculous leaps. I have spent a great deal of time in this book showing evidence that this is actually what has happened. Dawkins uses this chapter on punctuation to also get rid of such ideas of saltational leaps because they are not Darwinian. He gives that very interesting quote from Darwin, which I mentioned in Chapter 14 (writing to his friend Lyell about ideas of divine intervention):

'If I were convinced that I required such additions to the theory of Natural Selection, I would reject it as rubbish ... I would give nothing for the theory of Natural Selection, if it requires miraculous additions at any one stage of descent.'[26]

Here we see Darwin's 'scientism' most starkly. He is utterly wedded to a mechanistic scientific explanation for his theory. There is no room for God, not because there is any evidence against it but because it is simply unacceptable in his world-view. How like Dawkins this is. Dawkins is at least being a faithful disciple.

Dawkins then uses the example of the human embryo starting from a single cell and *gradually* becoming a complete person as a way of supporting the idea that we have evolved gradually from an amoeba-like single-celled creature. It is hard to believe that he could get this so wrong. The development of the human embryo is a process as different from evolution as chalk from cheese (actually more different). It is simply a non-argument. Instead, and more reasonably, I could use the amazing development of the human embryo as an argument against a purposeless random process such as evolution.

My personal opinion is that there is genuine punctuation in the development of life – there is stasis and sudden appearance of new species (similar but different to those before) and there are the sudden larger saltational leaps that we see – such as in the appearance of bacteria, the Cambrian explosion, the appearance of birds and the appearance of humans. This is what the evidence shows. If such evidence requires us to contemplate a divine input then any true scientist will allow it to do so.

Chapter 10. The one true tree of life

This chapter is about the classification of living things. It is obvious that organisms belong to definite groups. Aristotle classified animals into those with and without blood and made a good attempt at ordering the groups of animals that he knew into distinct classes. Before Darwin, taxonomists (experts in classifying living things) made many more attempts. An important one is that of M. Milne Edwards in 1844. Like most taxonomists, his classification of animals showed clearly demarcated groups arranged in a hierarchy. What we mean by hierarchy is that each animal group is part of a group, which is again part of a larger more inclusive group. Dawkins in this chapter gives a description of this here:

'Another way of representing this idea of strict hierarchy is in terms of 'perfect nesting'. We write the names of any set of animals on a large sheet of paper and draw rings round related sets. For example, rat and mouse would be united in a small ring indicating that they are close cousins, with a recent common ancestor. Guinea-pig and capybara would be united with each other in another smaller ring. The rat/mouse ring and the guinea-pig/capybara ring would in turn, be united with each other (and beavers and porcupines and squirrels and lots of other animals) in a larger ring labelled with its own name, rodents. Inner rings are said to be 'nested' inside larger, outer rings. Somewhere else on the paper, lion and tiger would be united with one another in a small ring. This ring would be included with others in a ring labelled cats. Cats, dogs, weasels, bears etc. would all be united, in a series of rings within rings, in a single large ring labelled carnivores. The rodent ring and the carnivore ring would then take part in a more global series of rings within rings in a very large ring labelled mammals.'[27]

Essentially what he is describing here is similar to what taxon-omists have done since Aristotle. The very interesting thing is that Darwinism has not changed the basic hierarchical structure of taxonomy. I will quote Dawkins again now because he writes about a very important feature of the system of rings described. They are 'perfectly nested'.

> 'The important thing about this system of rings within rings is that it is perfectly nested. Never, not on a single solitary occasion, will the rings that we draw intersect each other. Take any two overlapping rings, and it will always be true to say that one lies wholly inside the other. The area enclosed by the inner one is always totally enclosed by the outer one: there are never any partial overlaps.'[28]

He goes on to explain that this property of perfect nesting is not exhibited by classifications of say books in a library, lan-guages, soil types or schools of thought in philosophy. In all these cases there is definite overlap between different categories.

Dawkins states this truth without apparent embarrassment. His explanation is that in the evolving tree of life, once the tree has branched beyond a certain minimum distance (the species), the branches never come together again. This, to him, accounts for the utterly separate, non-overlapping nature of the rings in clas-sification.

This is extraordinary because we know from Darwinian theory that in fact the evolution of organisms is supposed to have hap-pened very gradually, step-by-step. If we were to see a perfect fos-sil record than we would expect to see a linear continuum with, instead of distinct cut-off points, a 'smeary' and blurred steady transition from primitive to modern animals. If we had those ani-mals all alive today we would see definite overlap – not distinct sets of rings that are present now. We would, in fact, see overlapping of the rings all the time. Why then is there absolutely no exception to the rule that the rings do not ever overlap? Dawkins himself proclaims this non-overlapping nature of 'perfect nesting' very clearly. Why, for example, do we see absolutely no overlapping between reptiles and mammals? The fact is that if they evolved in a smooth way from a common ancestor then there must have been a time when there was an in-between, transitional form which was neither reptile nor mammal. There is no particular reason why such a transitional form should not actually have survived to today. One should expect to have at least a few such transitional survivors in the whole of the animal kingdom with at least one case of 'non-perfect nesting'. But we do not and Darwinian theory cannot explain this.

The perfect nesting exhibited in classification of living things is, contrary to what Dawkins thinks, clear evidence against the gradualist Darwinian evolution that he proclaims. It is explained, however by a theory which involves saltations – those large macro leaps in development which I have discussed before. One would expect discontinuity and perfect nesting in a saltational model. This also fits very well with the actual fossil evidence of discontinuity which I described earlier.

It is hardly surprising, therefore, that within the discipline of taxonomy there is a degree of chaos and in-fighting. Dawkins describes very well the various categories of taxonomist and the extremely divergent views of the experts. Many taxonomists now do not believe that when classifying organisms there should be any attention paid to evolutionary relationships. This is for the admirable reason that one needs to be utterly objective and not swayed by anything other than the actual characteristics of the organism. Some, rejoicing in the name of 'transformed cladists', have gone as far as to say that there is no real way of finding any form of evolutionary relationship from observing the classification of organisms. This view, of course, is anathema to Dawkins and he spends some time attacking it. What we can say, though, is that the classification of living things shows up the weaknesses in the whole edifice of Darwinism. The discontinuities and 'perfect nesting' speak volumes to us, if we would but realise it.

Another area that he tackles in this chapter is the way in which we can now look at molecular sequences of proteins and DNA and compare these between different animal groups. I have already discussed this discipline of 'molecular phylogeny' in the chapter on genetics. He mentions a study that looked at five different protein molecules from eleven different animals to see if the five protein sequences would all agree on an evolutionary tree relationship between the animals. He reports that in all five protein sequences the computer agreed that humans, chimps and monkeys were close to each other but that there was no unanimity about the other eight animals. He explains away this lack of clarity about relationships between these eight animals rather too easily (basically invoking ideas of convergence and coincidence). He does not mention the extremely poor record of molecular phylogeny in general – such that, in molecular terms, the rabbit seems closer to primates than to rodents and sea urchins are grouped with chordates. I have already discussed in Chapter 2 how analysis of RNA does not show any clear founder organism for the tree of life.

It is well known that we share around 98 per cent of the same DNA gene sequences as chimps. This does seem confirmation of a relationship based on common descent. This is definitely not something that creationists can avoid. It suggests very strongly that we are related. The question is whether this clear relationship involves changes in gradual fashion in a Darwinian, smooth, sequence. The evidence is against this – as I have already explained, particularly in Chapter 6, the appearance of humanity looks far more like a saltational event. This does not in any way exclude a relationship with apes; it actually upholds the idea. A similar DNA profile is expected if the precursors of the saltated humans were apes.

I have previously mentioned the fact that other parts of the DNA, known as introns, have recently been found not to be redundant 'junk' but have definite roles within organisms – this may account for the degree of closeness of chimp and human DNA when only looking at the genes.

This chapter has highlighted for me the huge problems that Darwinian theory has in accounting for the classification of life – in particular the discrete, non-overlapping sets of organisms that we observe, lacking any sign of transitional overlaps that one should expect.

Chapter 11. Doomed rivals

Most of this chapter is concerned with de-bunking Lamarckism. I have no intention of discussing this because it is essentially a non-issue. Hardly anyone seriously believes in Larmarckian evolution.

He mentions the 'neutralist' ideas of the geneticist Kimura. Kimura has shown that many, if not most mutations (for instance, in the gene for the haemoglobin protein) are apparently neutral, meaning that they have neither a beneficial nor a detrimental effect. Dawkins has no quibble with this idea apparently but merely asserts that those other mutations that have real survival effect are the ones that drive evolution.

He attacks the ideas of those whom he calls 'mutationists'. I have already, in Chapter 9, mentioned both De Vriess and Bateson in the early twentieth century – naturalists who had little time for the power of selection but saw macro-mutations as being the force driving evolution. Dawkins sees this as anathema because of course we need to invoke some mystical directing to the macro-mutations if they are to produce anything complex or useful. Dawkins understandably attacks this but fails to recognise these men's insight that selection is actually quite inert and powerless to produce complexity. And, once again, he fails to recognise the fossil evidence for just such macro saltational leaps that De Vriess and Bateson described.

He then covers some other aberrant theories, which need not concern us here, and then, in a flourish, cannot resist an attack on creationism. His comments about the Genesis story are frankly offensive. His statement that it has no more status than any other myth, such as, 'the belief of a particular West African tribe that the world was created from the excrement of ants',[29] is unworthy of comment. I am not going to try to out-argue him about Genesis (it would take another book or two). I will argue with him (again), though, about his idea that God could not be an eternal, uncreated, complex being. He says:

'If we want to postulate a deity capable of engineering all the organised complexity in the world, either instantaneously or by guiding evolution, that deity must already have been vastly complex in the first place.'[30]

He then demands an explanation for this complex being of prodigious intelligence and is attempting to dismiss any idea of a pre-existent eternal God in one paragraph. Richard Dawkins clearly has an inflated idea of his competence in metaphysics. As I have stated before, he seems to be unaware that it did not take Darwinism to pose such questions about God. The concept of a 'first mover' has a long and respectable history in philosophy. Many philosophers down the ages have been forced to accept that to have anything at all in existence requires an original prime architect. The concept of a pre-existent all-powerful creator is one that is outside his field. He seems to compound his ignorance of this by assuming that his neat idea (that if God created us then we must invoke some idea of how he was created) is new. A pre-existent, uncreated God is something science cannot prove or disprove. It amazes me that Dawkins can pretend to do so here.

He finishes by summarising the argument for gradual evolution by cumulative selection, emphasising the gradual steps that we can conceive of in the emergence of the eye. He agrees that to get an eye from nothing in one generation, perfect and whole is beyond credulity and that the odds against it 'will keep us writing noughts till the end of time'.[31] In Chapter 9 on selection I have shown mathematically that, in fact, the series of steps that he requires to form the eye are no more probable when added together in a series. Selection, as always, is a red herring and has no part in the actual mutations required to get novel complexities.

Conclusion

I have found going through Dawkins' book both helpful and depressing. It has been helpful because it has thrown into light the huge inconsistencies that Darwinism has and the failure of its chief 'bulldog' to make any sort of good case for it. It has been depressing because this view holds such sway in the scientific community still despite the enormous weight of scientific evidence against it. I am comforted somewhat by Thomas Kuhn's observations that scientific paradigms are not set in stone but undergo shifts. There is enormous inertia in the Darwinian paradigm – but shift it will.

1. Richard Dawkins (1986). *The Blind Watchmaker.* Longman Scientific and Technical.
2. *Ibid.,* p. ix.
3. *Ibid.,* p. x.
4. *Ibid.,* p.xi
5. *Ibid.,* p.5.
6. *Ibid.,* p.6.
7. *Ibid.,* p.14.
8. *Ibid.,* p21.

9. *Ibid.*, P.38

10. Raymond and Lorna Coppinger, (2001). *Dogs. A New Understanding of Canine Origin, Behavior and Evolution.* Chicago University of Chicago Press.

11. Dawkins, *The Blind Watchmaker*, P.45

12. *Ibid.*, P. 50

13. *Ibid.*, p.77.

14. *Ibid.*, p.77

15. *Ibid.*, p.79

16. Ibid., P.P 85-86

17. *Ibid.*, p126.

18. *Ibid.*, P.140.

19. *Ibid.*, P.141.

20. *Ibid.*, P.141.

21. *Ibid.*, P.141.

22. *Ibid.*, p.169.

23. N. Eldredge and SJ. Gould, (1973). 'Punctuated Equilibria: An Alternative to Phyletic Gradualism.', in *Models in Paleobiology*, ed. TJM. Schopf. San Francisco: Freeman, Cooper and Co., pp. 82–115.

24. Stephen Jay Gould (2002). *The Structure of Evolutionary Theory.* Harvard: Belknap Press,

25. Lidgard and McKinney In Jackson (eds) (2001). *Evolutionary Patterns – growth, form and tempo in the fossil record.* Chicargo: University of Chicago Press.

26. F.Darwin (ed.) (1888). *The Life and Letters of Charles Darwin.*: John Murray.

27. Dawkins, *The Blind Watchmaker*, P.259

28. Ibid., P.259

29. *Ibid.*, p.316.

30. *Ibid.*, p.316

31. *Ibid*

Chapter Sixteen

Conclusions

I began this investigation unsure of what I would find in the way of evidence for or against Darwinism. It has been a fascinating journey, which is continuing. Am I clearer in my views of what has happened to account for life on Earth?

In one sense I am not – there remains deep mystery. We simply do not know how the first replicating cells arose and we have no idea how random events could have thrown up the fantastic complexity that followed. If there is one thing that I hope the reader will take on board it is the absolute lack of any clear model available to science which can explain the existence of life. Darwinism has been shown to be a totally inadequate explanation – apart from at the micro-evolutionary level.

Nothing that I have found is radically new; I merely gathered the thoughts and studies of others. It has not been easy to do so in the face of the juggernaut of neo-Darwinist thinking, but the evidence is available to us all. In fact it has been said and stated many times in past years – not least by the great Scottish geologist and palaeontologist Hugh Miller at the time of Darwin.[1] The scientific community has, in general, turned a deaf ear but the paradigm is creaking.

The concept of shifting paradigms in science has been well expounded by Thomas Kuhn. His book The Structure of Scientific Revolutions, written in 1970, opened my eyes to the way in which scientific theories may seem to be set in stone but are sometimes much more shaky than the proponents can fathom.[2]

Ptolemy's model of the known universe provides a good example. His idea of heavenly spheres of ever greater complexity, rotating around the earth, was accepted for centuries; not because the ancients were ignorant (they were, in fact, superb astronomers) but because the model of the spheres provided a very reasonable system to account for the movement of the stars and planets. Cracks began to show in the theory when the detailed movements of the planets could not be accounted for adequately. This signalled the end and the eventual replacement of the theory with a more correct model. The scientists of the days of Ptolemy were no fools and they observed very carefully. The fact that their system was wrong was not their fault. Their system provided a very reasonable way of looking at the universe. The observed facts of stars and planets seemed to fit in with it.

Darwinism is very similar to this. We are taught the theory in school and we take it on trust. The entire scientific establishment seems behind it. Careers in science depend on it. To criticise it as an academic is to risk professional isolation and demotion. There is, though, an increasing, evidence-based momentum within the heart of the scientific community showing up the fault lines in the orthodoxy. I have quoted a number of established academics who are questioning the 'ism' that Darwin began. The word saltation is even on the lips of some.

The situation is not helped by those in the creationist community who are not just anti-Darwin but seem anti-science as well. To say that the Earth is only a few thousand years old is unhelpful. Sincere people are making such assertions but they do a disservice to the pursuit of truth and they give an extremely bad name to the movement that is growing to displace Darwinism. That I still uphold Genesis to be true is too much to explain in this book.

Critics will assert that I have merely twisted facts and ideas to fit in with my belief in a creator. Of course, this can be a temptation. I have, however, tried very hard not to do so and to look as objectively as possible at the available evidence. The fact is that we all start with some sort of belief system – be it atheist, deist, theist, pantheist or agnostic. It can be hard not to allow such beliefs to colour one's conclusions at the expense of accuracy. Readers will have to judge whether I have succeeded or not.

And so, at the end of this stage of the journey, which I began some years ago, I have come to certain conclusions. All the evidence that I have looked at has led me to believe that life has been designed and that all the major innovations have occurred in saltations, the like of which no Darwinist can explain. The evidence before us shows that a creator has been active throughout.

William Dembski, the philosopher, mathematician and theologian wrote that those who argue for intelligent design in nature say:

> 'that Darwinism is on its own terms a failed scientific research program – that it does not constitute a well-supported scientific theory, that its explanatory power is severely limited, and that it fails abysmally when it tries to account for the grand sweep of natural history.'[3]

The poem by Gerard Manley Hopkins at the beginning of this book shows, quite brilliantly, what most of us at some time feel deep in our bones; in those moments of clarity when nature speaks to us of some majesty which upholds and produces its wonders. Modern western man is in danger of losing this even though it is taken for granted in other cultures and was always understood by his forebears. Let us not allow a very flawed theory to deprive us of such insight.

1. Hugh Miller (1857). *The Testimony of the Rocks.* 2001 edn. St Matthew Publishing Ltd.
2. TS. Kuhn (1970). *The Structure of Scientific Revolutions.* Chicago: University of Chicago Press.
3. William A. Dembski (1998). *Mere Creation. Science. Faith and Intelligent Design.* Inter Varsity Press.

Bibliography

Bateson, W. (1909). 'Heredity and Variation in Modern Lights', in *Darwin and Modern Science*. Cambridge University Press.

Behe, Michael (1996). *Darwin's Black Box*. Free Press.

Benton, Michael (1997). *Vertebrate Palaeontology*. Chapman and Hall.

Benton, M. J. and Hitchin, R. (1996). 'Testing the quality of the fossil record by groups and by major habitats', *Historical Biology*, 12, 111–57.

Bernal, J. D. (1967). *The Origin of Life*. Cleveland, OH: World Publishing Co.

Brunet, Michel *et al.* (2002). 'A new hominid from the upper Miocene of Chad, Central Africa', *Nature*, 418,pp 145–51.

Clack, Jennifer (2002). *Gaining Ground. The Origin and Evolution of Tetrapods*. Indiana: Indiana University Press.

Clarkson, E. N. K (1993). *Invertebrate Palaeontology and Evolution*. 3rd edn. Chapman and Hall.

Conway Morris, Simon (1998). *The Crucible of Creation*. The Burgess Shale and the rise of animals. Oxford University Press.

Conway Morris, Simon (2002). 'Body Plans', in *Encyclopedia of Evolution, vol 1.* Oxford University Press.

Conway Morris, Simon (2003). *Life's Solution. Inevitable Humans in a Lonely Universe.* Cambridge University Press.

Coppinger, Raymond and Lorna (2001). *Dogs. A New Understanding of Canine Origin, Behavior and Evolution.* Chicago: University of Chicago Press.

Crow, T. J. (ed.) (2002). *The Speciation of Modern Homo sapiens. Proceedings of the British Academy.* Oxford: Oxford University press.

Currie, Philip J. (2004). *Feathered Dragons. Studies on the Transition from Dinosaurs to Birds.* Indiana: Indiana University Press.

Darwin, Charles (1859). *The Origin of Species.* John Murray.

Darwin, Charles (1871). *The Descent of Man and Selection in Relation to Sex.* Princeton: Princeton University Press.

Darwin, Charles (1997 edn). *The Voyage of the Beagle.* Wordsworth Editions.

Darwin, Francis (ed) (1888). *The Life and Letters of Charles Darwin.* John Murray.

Davies, Paul (1998). *The Fifth Miracle.* Allen Lane: Penguin

Dawkins, Richard (1984). *The Selfish Gene* (new edn). Oxford University Press.

Dawkins, Richard (1986). *The Blind Watchmaker.* Longman Scientific and Technical.

Dawkins, Richard (1995). *River out of Eden.* London: Weidenfeld & Nicolson.

De Beer, Gavin (1971). *Homology: An Unsolved Problem.* London. Oxford University Press

De Vriess, H. (1909). *The Mutation Theory.* Appleton & Co.

Dembski, William A. (ed.) (1998). *Mere Creation. Science, Faith and Intelligent Design.* Inter Varsity Press.

Denton, Michael (1986). *Evolution, a Theory in Crisis.* Adler & Adler.

Desmond, Adrian and Moore, James (1991). *Darwin.* Penguin.

Eldredge, N. and Gould, S. J. (1973). 'Punctuated Equilibria: An Alternative to Phyletic Gradualism', in *Models in Paleobiology,* ed. T. J. M. Schopf. San Francisco: Freeman, Cooper and Co., pp. 82–115.

Fortey, Richard, (1998.) *Life – An Unauthorised Biography.* Flamingo.

Gee, Henry (2000). *Deep Time.* Fourth Estate.

Gould, Stephen Jay (1990). *Wonderful Life. The Burgess Shale and the Nature of History.* New York: Vintage.

Gould, Stephen Jay (2002). *The Structure of Evolutionary Theory.* Harvard: Belknap Press.

Greenfield, Susan (1997). *The Human Brain.* Phoenix.

Griffiths, Miller, Suzuki, Lewontin and Gilbart (1999). *An Introduction to Genetic Analysis.* W. H. Freeman.

Harrison, E. (1985). *Masks of the Universe.* New York: Collier Books, Macmillan.

Hunter, Cornelius (2001). *Darwin's God.* Brazos Press.

Jackson, J. B. C., Lidgard, S. and McKinney, F.(2001). *Evolutionary Patterns. Growth, Form and Tempo in the Fossil Record.* Chicago Universiy of Chicago Press.

Johnson, Phillip (1991). *Darwin on Trial.* Washington DC: Regnery Gateway.

Jones, Steve (1999). *Almost Like a Whale.* Anchor.

Joyce, G. F. (1989). 'RNA evolution and the origins of life', *Nature*, 338, pp.217-23.

Lewin, Roger (1996). *Patterns in Evolution. The New Molecular View. Scientific American* Library.

Lewin, Roger (1998). Much material is from The Origin of Modern Humans. *Scientific American* Library.

McCormick, T. and Fortey, R. (2002) 'The Ordovician Trilobite Carolinites, a test case for microevolution in a macrofossil lineage', *Palaeontology.* vol.45, part 2.

Miller, Hugh (1857). *The Testimony of the Rocks.* (2001 edn). St Matthew Publishing Ltd.

Moore, Janet (2001). *An Introduction to the Invertebrates.* Cambridge University Press.

Muller, F. M. (1996). 'Lectures on Mr Darwin's philosophy of language.' in: *The Origins of Language* (ed. R. Harris). Bristol: Thoemmes Press, pp.147–233.

Norell, M. A. and Novacek, M. J. (1992). 'The fossil record and evolution: comparing cladistic and palaeontological evidence for vertebrate history', *Science*, 255, pp1690–3.

O'Hear, Anthony (1997). *Beyond Evolution. Human nature and the limits of evolutionary explanation.* Oxford: Clarendon Press.

Parker, Andrew (2003). *In the Blink of an Eye: the cause of the most dramatic event in the history of life.* Free Press.

Pinker, S. (1994). *The Language Instinct.* London: Allen Lane.

Popper, Karl (1934). 'Scientific Method', in *Popper Selections.* Princeton: Princeton University Press.

Rattray Taylor, Gordon (1983). *The Great Evolution Mystery.* Secker & Warburg.

Rees, Martin (2002). *Our Cosmic Habitat.* London: Weidenfeld & Nicolson.

Ridley, Mark (1996a) *Evolution.* Blackwell Science.

Ridley, Matt (1996b). *The Origins of Virtue.* Penguin.

Rowe, A. W. (1899). 'An analysis of the genus *Micraster*, as determined by rigid zonal collecting from the zone of *Rhynchonella cuvieri* to that of *Micraster coranguinum.*' *Quarterly Journal of the Geological Society of London* 55, pp.494–547.

Ruse, Michael (1999a). *The Darwinian Revolution*. Chicago.

Ruse, Michael (1999b). *The Darwinian Revolution. Science Red in Tooth and Claw*. Chicago: University of Chicago Press.

Ruse, M. and Wilson, E. O. (1986). 'Moral philosophy as applied science', *Philosophy* 61, pp. 173–92.

Sampson, Philip (2000). *Six Modern Myths*. Inter Varsity Press.

Schopf, J. William (2001). *Cradle of Life: The discovery of Earth's earliest fossils*. Princeton: Princeton University. Press.

Shubin, *et al.* (1997). 'Fossils, genes and the evolution of animal limbs', *Nature* 388 (6643). PP.639–648.

Skelton, Peter (1993). *Evolution. A biological and palaeontological approach*. Addison Wesley with Open University.

Walsh, John E. (1996). *Unravelling Piltdown*. Random House.

Wells, Jonathan (1998). *Unseating Naturalism. In 'Mere Creation'*. Science, Faith and Intelegent Design (*Ed. W. Dembski.*) Inter Varsity Press.

Wells, Jonathan (2000) *Icons of Evolution. Science or Myth?* Regnery Publishing.

Wickramasinghe, C. (2001). *Cosmic Dragons. Life and Death on our Planet*. Souvenir Press.

Williams, Peter S. (2002). *Intelligent Design, Aesthetics and Design Arguments*. Available at http://www.iscid.org/boards/ubb-get_topic&-f-10&-t-000014.html

Woesse, Carl (1998). *'Proceedings of the Natural Academy of Sciences.'* *USA*, 95, 6854.

Wolfe, Stephen (1993). *Molecular and Cellular Biology.* Belmont, CA: Wadsworth Publishing Co.

Wong, Kate (2003). 'An ancestor to call our own. New look at human evolution', *Scientific American*, – special edition.

Wray, Gregory A and Abouheif, Ehab (1998). 'When is homology not homology?', *Current Opinion in Genetics and Development* 8, pp 675–80.